Risk Assessment and Risk Management

University
Of Dundee
UNIVERSITY LIBRARY

School of Nursing and Midwifery
Fife Campus

Date of Return

ISSUES IN ENVIRONMENTAL SCIENCE AND TECHNOLOGY

TITLES IN THE SERIES:

FORTHCOMING:

How to obtain future titles on publication

A subscription is available for this series. This will bring delivery of each new volume immediately upon publication. For further information, please write to:

The Royal Society of Chemistry
Turpin Distribution Services Limited
Blackhorse Road
Letchworth
Herts SG6 1HN, UK

Telephone: +44 (0) 1462 672555
Fax: +44 (0) 1462 480947

ISSUES IN ENVIRONMENTAL SCIENCE
AND TECHNOLOGY

EDITORS: R. E. HESTER AND R. M. HARRISON

9

Risk Assessment and Risk Management

THE ROYAL
SOCIETY OF
CHEMISTRY
Information
Services

**Dedicated to the memory of the late
Professor Roger Perry**

ISBN 0-85404-240-7
ISSN 1350-7583

A catalogue record for this book is available from the British Library

Published by The Royal Society of Chemistry, Thomas Graham House,
Science Park, Milton Road, Cambridge CB4 4WF, UK
For further information see our web site at www.rsc.org

Typeset in Great Britain by Vision Typesetting, Manchester
Printed and bound by Redwood Books Ltd., Trowbridge, Wiltshire

Preface

There has been much debate recently about risk assessment, considered by many analysts to be an objective scientific tool, as being variously influenced by the broader ethical, social, political, economic and institutional issues which characterize risk management. The impact of such influences and the extent to which they occur have important practical implications both for risk assessors and for decision makers. This volume examines a range of practical applications of risk assessment methods and risk management procedures in the broad context of environmental science and technology. These serve to illustrate and to clarify the concepts and terms which are widely used by practitioners in the field as well as giving deep insights into the practical applications of risk assessment.

This volume begins with an article by Simon Gerrard of the University of East Anglia and Judith Petts of Loughborough University which immediately demonstrates the need for careful analysis of current practices and their consequences. It opens with a succinct overview of bovine spongiform encephalopathy (BSE), which has proved to be one of the most controversial risk management issues of the decade. It goes on to set the scene with a consideration of the contrasting roles of logical positivism and cultural relativism and the mixture of facts and value judgements that tends to characterize debate in this area. The blurred boundaries between risk assessment, risk evaluation, risk analysis and risk management are examined and illustrated within decision-making processes relating to the siting of waste incinerators and the management of contaminated land in both the UK and the USA.

The way in which growing public demand for safer and healthier workplaces, cleaner environment, better housing and safer food is putting pressure on government departments to examine the way they regulate risks is analysed by Jim McQuaid and Jean-Marie Le Guen of the UK Health and Safety Executive (HSE) in the second article. This reviews the use of risk assessment in government and sets out the criteria developed by the HSE for assessing the tolerability of risk. Simon Halfacree of the UK Environment Agency then illustrates the work of this other government agency with particular reference to water pollution and the associated risks. A centrepiece of this is the use of computer modelling as an aid to risk assessment in the case of accidental contamination of rivers and estuaries.

The practice of quantitative cancer risk assessment related to carcinogens in the environment is examined by Steve Hrudey of the University of Alberta. He

v

provides a critical evaluation of regulatory practice in this area and gives us insights into several of the major controversies and misunderstandings which have arisen. Risks to human health also are the subject of the article by Gev Eduljee of the Environmental Resources Management company. He considers risks associated with the widespread practice of landfilling of household wastes. The need to develop a sustainable and integrated approach to waste management which minimizes risk to human health is felt by essentially all communities and countries. This article provides a thorough and detailed examination of the issues and uses a quantitative case study to illustrate methods of risk evaluation and control.

Environmental risk assessment and management of chemicals is the subject of the article by Derek Brown, who is an independent consultant. His emphasis is on European Union legislation, but the issues raised have international significance. Detergent surfactant biodegradability provides a vehicle for the examination of various test methods and risk assessment factors leading to appropriate management strategies. The final article, by Steve Maund of Zeneca Agrochemicals and Neil Mackay of JSC International Ltd, reviews approaches to aquatic risk assessment and management for pesticides. This provides another view of some of the methods described in the preceding article and identifies effect concentrations (from toxicity studies) for comparison with predicted environmental concentrations. Where further evaluation of potential risks is required the article describes higher tier effects studies, such as field studies or exposure modelling, and the mitigation of risks by modifying the pesticide use patterns.

Taken together, this set of articles provides a detailed and wide-ranging review of the many aspects of risk assessment and risk management which have excited so much debate and controversy in recent times. It illustrates a variety of practical applications and exposes the limitations and uncertainties that are intrinsic to the subject. The volume as a whole should be essential reading for all those involved in the assessment and management of risk, particularly but not exclusively in the context of environmental science and technology.

Professor Roger Perry, formerly a member of the Editorial Advisory Board of Issues in Environmental Science and Technology, died in October 1995. Professor Perry had a deep involvement in the scientific and policy aspects of environmental risk assessment and management, and his forceful and often controversial contributions to the debates on environmental policy issues were sometimes highly influential. We are pleased to dedicate this volume to his memory.

Ronald E. Hester
Roy M. Harrison

Contents

Contents

Editors

Ronald E. Hester, BSc, DSc(London), PhD(Cornell), FRSC, CChem

Ronald E. Hester is Professor of Chemistry in the University of York. He was for short periods a research fellow in Cambridge and an assistant professor at Cornell before being appointed to a lectureship in chemistry in York in 1965. He has been a full professor in York since 1983. His more than 300 publications are mainly in the area of vibrational spectroscopy, latterly focusing on time-resolved studies of photoreaction intermediates and on biomolecular systems in solution. He is active in environmental chemistry and is a founder member and former chairman of the Environment Group of The Royal Society of Chemistry and editor of 'Industry and the Environment in Perspective' (RSC, 1983) and 'Understanding Our Environment' (RSC, 1986). As a member of the Council of the UK Science and Engineering Research Council and several of its sub-committees, panels and boards, he has been heavily involved in national science policy and administration. He was, from 1991–93, a member of the UK Department of the Environment Advisory Committee on Hazardous Substances and is currently a member of the Publications and Information Board of The Royal Society of Chemistry.

Roy M. Harrison, BSc, PhD, DSc (Birmingham), FRSC, CChem, FRMetS, FRSH

Roy M. Harrison is Queen Elizabeth II Birmingham Centenary Professor of Environmental Health in the University of Birmingham. He was previously Lecturer in Environmental Sciences at the University of Lancaster and Reader and Director of the Institute of Aerosol Science at the University of Essex. His more than 250 publications are mainly in the field of environmental chemistry, although his current work includes studies of human health impacts of atmospheric pollutants as well as research into the chemistry of pollution phenomena. He is a past Chairman of the Environment Group of The Royal Society of Chemistry for whom he has edited 'Pollution: Causes, Effects and Control' (RSC, 1983; Third Edition, 1996) and 'Understanding our Environment: An Introduction to Environmental Chemistry and Pollution' (RSC, Second Edition, 1992). He has a close interest in scientific and policy aspects of air pollution, having been Chairman of the Department of Environment Quality of Urban Air Review Group as well as currently being a member of the DETR Expert Panel on Air Quality Standards and Photochemical Oxidants Review Group, the Department of Health Committee on the Medical Effects of Air Pollutants and Chair of the DETR Atmospheric Particles Expert Group.

Contributors

D. Brown, *Consultant, 26 Brunel Road, Paignton, Devon TQ4 6HN, UK*

G. Eduljee, *Environmental Resources Management, Eaton House, Wallbrook Court, North Hinksey Lane, Oxford OX2 0QS, UK*

S. Gerrard, *Centre for Environmental and Risk Management, School of Environmental Sciences, University of East Anglia, Norwich NR4 7TJ, UK*

S. Halfacree, *Environment Agency, Rivers House, St. Mellons Business Park, St. Mellons, Cardiff CF3 0LT, UK*

S. Hrudey, *Department of Public Health Sciences, Faculty of Medicine, University of Alberta, Edmonton, Alberta, Canada T6G 2G3*

J.-M. Le Guen, *Health and Safety Executive, Rose Court, 2 Southwark Bridge, London SE1 9HS, UK*

N. Mackay, *JSC International Ltd., Osborne House, 20 Victoria Avenue, Harrogate, North Yorkshire HG1 5QY, UK*

J. McQuaid, *Health and Safety Executive, Rose Court, 2 Southwark Bridge, London SE1 9HS, UK*

S. Maund, *Zeneca Agrochemicals, Jealott's Hill Research Station, Bracknell, Berkshire RG42 6ET, UK*

J. Petts, *Centre for Hazard and Risk Management, Loughborough University, Loughborough, Leicestershire LE11 3TU, UK*

Isolation or Integration?
The Relationship Between Risk Assessment and Risk Management

SIMON GERRARD AND JUDITH PETTS

1 Introduction

The potential impact of bovine spongiform encephalopathy (BSE) has proved to be one of the most controversial risk management issues of the decade. Whilst interspersed with numerous other risk controversies, BSE has outperformed many of its contemporaries in terms of length, scientific uncertainty, economic impact and public outrage. One of the most pressing decision problems was how best to deal with the carcasses such that the public would be confident that the threat of BSE contamination was negligible. In June 1997 the Environment Agency in the UK issued a press release explaining the results of trial burns it had conducted at two coal-fired power stations.[1] It reported that:

'The risk of human infection resulting from burning cattle cull wastes in power stations would be negligible. A detailed risk assessment, carried out by the Agency, based on test rig trial burning of meat and bonemeal (MBM) and tallow from cattle slaughtered under the Goverment's Over Thirty Month Scheme (OTMS) shows that the risk of an individual contracting CJD (*Creutzfeldt Jacob Disease*) would be as low as 1 in 30 000 million. This is 3000 times less than the risk of death by lightning'.

The press release went on to confirm that, though the studies showed that the risk to both the public and the workers would be negligible, the Agency would not be giving blanket approval to applications for cattle incineration. Approval would only be granted if all statutory requirements were met: 'Every application will be rigorously assessed on its own merits and there will also be widespread public consultation before a decision is taken'.

It is indicative of current risk debates that a regulatory agency chooses to announce the result of its quantitative risk assessment in the same breath as its assurance that, despite the extremely low risk, widespread public consultation

---

[1] *Burning Cattle Cull Waste In Power Stations—Environment Agency Says Risks Are Negligible*, Press Release Number 063/97, Environment Agency, Bristol, 25 June 1997.

1

will occur in advance of any decision. This example could be interpreted in different ways. Some may question why widespread consultation is necessary if the risk is so low. Others may ask what is so wrong with the risk assessment that widespread consultation is still necessary?

The rise of risk assessment as a tool for decision makers comes in the face of mounting criticism from those in industry who argue that it is too conservative in its assumptions and thus creates public fear and unnecessary financial hardship *and* from environmental and community groups who see the tool as too simplistic and narrow to deal with the complex reality of risk issues. Characterizing risk debates is fraught with difficulty as terms are interchangeable and possess different meanings for different parties. However, these positions may be characterized by the extent to which risk assessment is regarded merely as one of many tools available to decision makers or as a decision-making process itself.

Risk management encompasses disciplines from the natural, engineering, political, economic and social sciences. One of the key issues highlighted by the multidisciplinarity of risk management is whether risk assessment as a scientific process can and should be separated from risk management. The basis of the arguments for and against separation are rooted in fundamental views of the role of science and society. Over the 20–30 year period of risk management research much has been learned, though many of the important lessons (not least stakeholder involvement) relate as much to the challenges of working in a multidisciplinary setting as to the development of the single disciplines involved.

Whilst we recognize that risk assessment is clearly a part of the process of managing risks, we also note that there are many different risk assessment approaches in different decision-making contexts. It is beyond the scope of this article to discuss all of the variabilities. Instead, we focus on example applications of risk assessment at different tiers of decision making: project/site-specific, strategic regulation and policy making. This includes two distinct types of risk assessment. Quantitative risk assessment relates to an activity or substance and attempts to quantify the probability of adverse effects due to exposure. In contrast, comparative risk assessment is a procedure used for ranking risk issues by their severity in order to prioritize and justify resource allocation. The examples of these two types discussed here provide a means of illustrating (i) how risk assessment is being used in decision making, (ii) the issues which this use raises in relation to robustness, efficiency and fairness, and (iii) whether and how risk assessment can be more effectively integrated into risk management. First, however, it is important to provide some historical background to the isolation *versus* integration debate and to expand upon our definitions.

2 Historical Review

At a simple level the arguments for isolation or integration of risk assessment (particularly quantified risk assessment) into risk management can be reduced to the degree to which one believes that science and the scientific process can be regarded as wholly objective. Logical positivism regards science as objective, taking the view that scientific assessments of risk should be kept separate from the social and political aspects of decision making. Science is devoted to the establishment of fact and is therefore necessarily isolated from the value judgements that pervade the rest of the decision-making process. By contrast, cultural relativism argues that science and the scientific process are inextricably linked to subjective value judgements; that science is bound-up with political and social institutions and is thus unable to be wholly objective. Between the two positions of complete isolation and total integration exist a variety of other positions that base their arguments to a greater or lesser extent on the two endpoints. Of these, perhaps the most persuasive is the notion of scientific proceduralism,[2] which seeks to tread a somewhat cautious path between the two extremes. Scientific proceduralism focuses on the process by which scientific activities are conducted. It recognizes explicitly that science is not wholly objective and that subjective value judgements within technical risk assessments have to be acknowledged and dealt with in an appropriate manner. One of the strengths of this approach is that it does not argue for a wholesale rejection of risk assessment. Rather it focuses upon a blend of robust scientific and technical analysis, effective communication and stakeholder involvement.[3,4] This might seem like a compromise solution. There is little doubt that it is not an easy solution, not least in terms of how existing decision-making structures can adapt to meet the requirements of integration. However, the 1997 publication of a United States Presidential Commission report on risk assessment and risk management[5] not only suggests that the message of scientific proceduralism is now firmly on the political agenda, but also helps to put in context the development of the arguments.

At the beginning of the 1980s an increasing level of concern was expressed in the US that the scientific aspects of risk assessment were being corrupted by extraneous and irrelevant social policy dimensions. One high-level study proposed a return to the separation of facts and values[6] in the management of risks. The proposed scheme relied upon a tripartite system of scientific risk assessment, risk assessment policy and risk management. The first dealt with facts, the last with values, leaving risk assessment policy to liaise between the two. The point to make is that eventually science and policy *must* interact. The bridge between the two extremes, here called risk assessment policy, is constructed from decision rules based either on fact or on value, whichever is chosen as being a legitimate matter of policy choice. That science might somehow be conducted in isolation and occasionally deliver objective information, which the policy makers can then choose either to accept or reject, is not a plausible vision. In considering the nature of quantitative risk assessment, the US National Academy of Sciences

---

[2] K.S. Schrader Frechette, *Risk and Rationality. Philosophical Foundations for Populist Reforms*, University of California Press, Berkeley, CA, 1991, ch. 3, p. 29.

[3] T. O'Riordan, in *Innovation and Environmental Risk*, ed. L. Roberts and A. Weale, Belhaven Press, London, 1991, p. 149.

[4] T.C. Earle and G.T. Cvetkovich, *Social Trust: Towards a Cosmopolitan Society*, Praeger, Westport, CT, 1995.

[5] Presidential/Congressional Commission on Risk Assessment and Risk Management, *Framework for Environmental and Health Risk Management*, National Academy of Sciences, Washington, DC, 1997, vol. 1.

[6] National Research Council, *Risk Assessment in the Federal Government: Managing the Process*, National Academy Press, Washington, DC, 1983.

identified possibly 50 opportunities where scientists may have to make discretionary judgements, ranging from the kinds of hazards to study to the identification of most relevant exposure pathways.[7] This is supported by a number of studies that have investigated different teams' efforts to conduct risk assessments for the same process where, typically, risk estimates can vary by at least one order of magnitude, depending on the assumptions made at the outset.[8] Though opponents use these studies[9,10] to decry the use of risk assessment, we adopt a more moderate position which recognizes that science, and with it risk assessment, has to be directed. In this context, risk assessment is a tool which, if . used appropriately, can aid decision makers.

In the UK a similar discomfort has been felt by some scientists concerning the gradual incursion into their domain of social and political dimensions. An illustration of this can be found in the preface to the Royal Society's second risk management volume, published nine years after its initial report on the techniques of risk assessment.[11] The second report focused more broadly on risk management and included a chapter on risk perception and communication.[12] The preface to the report contained the explanation that the content should be viewed as a report not of the Society's but of the chapter authors' views. By disguising its inability to address the issue of how best to manage scientific and non-scientific material in terms of not wanting to 'pre-empt the very debate that the Council and the contributors wish to encourage', the Society left open the whole issue of the increasingly unhappy marriage of facts and values.

The search for truth championed by positivist philosophies in the development of knowledge within the natural sciences, in particular Popper's logical positivism, has dominated almost to the exclusion of alternative rationalities.[13] For example, in the risk field it remains a widely held belief that an actual risk number can be calculated with ever increasing precision for any particular technology, event or activity. The dominance of this prevailing attitude, particularly amongst decision makers new to the concept of risk, serves to promote the importance of technical assessments that subsequently drive the framing of hazard and risk management problems. It has been argued, not least by social scientists themselves, that social anthropology, history, economics, political science, cultural theory and the other social sciences have failed, to a large extent, to provide the same kind of explanatory power as the natural sciences.[14]

Risk perception and communication discussions were founded in the genuine expert puzzlement over active public opposition to technologies that scientists

[7] National Research Council, *Risk Assessment in the Federal Government. Managing the Process*, National Research Council, Washington, DC, 1983, p. 28.

[8] R. R. Kuehn, *Univ. Illinois Law Rev.*, 1996, **1**, 103.

[9] A. Amendola, *Nucl. Eng. Design*, 1986, **93**, 215.

[10] R. Ginsberg, *New Solutions*, 1993, winter, 8.

[11] The Royal Society, *Risk Assessment*, The Royal Society, London, 1986.

[12] The Royal Society, *Risk: Analysis, Perception and Management*, Report of a Royal Society Study Group, Royal Society, London, 1992.

[13] K. Popper, *Conjectures and Refutations: The Growth of Scientific Knowledge*, Harper Torchbooks, New York, 1965.

[14] A. Giddens, *New Rules of Sociological Method: A Positive Critique of Interpretative Sociologies*, Polity Press, Cambridge, 2nd edn., 1993.

thought were safe, with a belief that the public were irrational and merely had to be educated. However, following early work to identify what was an acceptable level of risk it became clear that the concept of risk means more to individuals than measurable fatalities or injuries. A broad range of qualitative characteristics has been identified by psychometric research, related not only to the nature of the potential harm, but also to the potential for control, the extent to which risk management institutions can be trusted to manage risks and concern over equity in risk bearing.[15-17] The socio-cultural literature has explained the reasons for such responses as a reflection of the social arrangements or institutions which people identify with or participate in.[18]

The development of risk communication research and discussion has mirrored, but also built upon, perception research, and provides the strongest support for an integration argument. Fischhoff[19] provides an historical review which charts the progress from a view of communication as a one-way process of educating people about acceptable risks, through recognition that communication has to be a two-way process, to current discussion of how to involve different stakeholders and particularly the public as partners in decision making. The risk communication literature supports the scientific procedural approach by allowing a greater degree of flexibility to be built into the social system through the adoption of decision-making systems which rely upon the rigours of positivistic science *and* upon the traditional (everyday) experiences of other social groups.

The extent of development of a scientific argument in favour of integration can be seen in the Presidential Commission report.[5] A new framework for risk management is presented which stresses (i) the engagement of stakeholders as active partners, (ii) an iterative approach which allows for new information to be built into the management process at any stages, and (iii) the need for health and environmental risks to be evaluated in their broader context rather than focused on single chemicals in single media. How far this report will meet favour with those rooted in logical positivism remains to be seen. In the UK, the debate is still largely open. A recent review of risk assessment in government departments recognizes the importance of stakeholder concerns and communication but still relegates solutions to continuing scientific/public controversies to those giving the public 'balanced information' about risks and 'enough time to reflect upon it'.[20] Meanwhile, a report for Parliamentary members in reviewing the isolation *versus* integration debate concludes in favour of the latter, stressing 'decision-making should be open, accountable and inclusive: seeking to achieve consensus; and

[15] B. Fischhoff, S. Lichtenstein, P. Slovic, S. L. Derby and R. L. Keeney, *Acceptable Risk*, Cambridge University Press, Cambridge, 1981.
[16] *Judgement under Uncertainty: Heuristics and Biases*, ed. D. Kahnemann, P. Slovic and A. Tversky, Cambridge University Press, Cambridge, 1982.
[17] P. Slovic, *Science*, 1987, **236**, 280.
[18] N. Pidgeon, C. Hood, D. Jones, B. Turner and R. Gibson, in *Risk: Analysis, Perception and Management*, Royal Society, London, 1992, p. 89.
[19] B. Fischhoff, *Risk Anal.*, 1996, **15** (2), 137.
[20] Health and Safety Executive, *Use of Risk Assessment in Government Departments*, Health and Safety Executive, Sudbury, 1996, p. 37.

Figure 1 The risk
management cycle

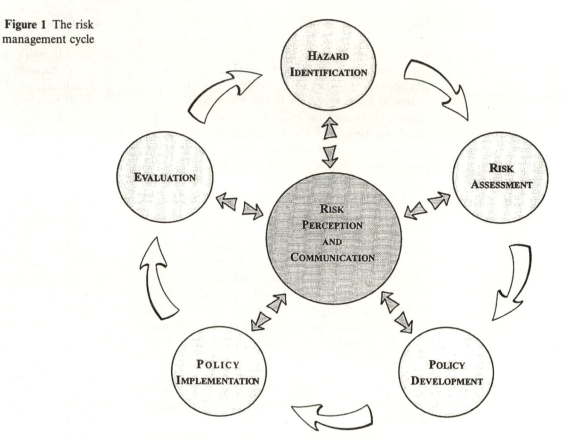

taking proper account of both the natural and social scientific dimensions of risk'.[21]

There remains no widely accepted, standard definition of risk assessment and management. The risk field is littered with alternative phrases which confuse discussions. The boundaries between risk assessment, risk evaluation, risk analysis and risk management are often blurred, leading to complications when discussing the relative importance of each, and confusion over the precise activities that comprise each component.

Many different schema have been developed to depict the structure and process of risk management. Traditionally, primarily linear approaches to understanding risk management have prevailed, reflecting the separation of risk assessment from risk management activities. However, we prefer to adopt a cyclical view of risk management as illustrated in Figure 1.[22] This differs from many of its counterparts in that the cycle emphasizes the importance of feedback to the extent that the starting and finishing points for risk management merge.

[21] Parliamentary Office of Science and Technology, *Safety in Numbers?—Risk Assessment in Environmental Protection*, POST, London, 1996, p. 53.

[22] B. A. Soby, A. C. D. Simpson and D. P. Ives, *Integrating Public and Scientific Judgements into a Tool Kit for Managing Food-related Risks, Stage I: Literature Review and Feasibility Study*, CERM Research Report No. 16, University of East Anglia, Norwich, 1993.

One effect of blurring these distinctions is to challenge the basis of science as a driving force and framing mechanism for solving risk management problems. To assist this shift in dominance further, at the heart of the risk management cycle lie the combined components of risk perception and communication, which reflect the fact that many of the examples of both good and bad practice in risk management hinge around the way in which debates are staged, competing views of risk are addressed and debated and, ultimately, decisions are made. Figure 1 is now close to that proposed by the Presidential Commission[5] and the UK Parliamentary Office of Science and Technology.[21]

Risk *management* cannot and should not be based simply on scientific understandings of how best to manage risks.[23-25] The incorporation of social preferences, often characterized in terms of the moral and ethical implications that lie at the heart of a democratic society, has to be addressed at some point within the decision-making process. The crucial question is at what stage and how should the scientific and technical assessment be integrated into risk management. Establishing some level of common ground here is still proving difficult as competing interests may gain or lose if the balance of power is altered.

3 Risk Assessment in Siting Decisions

This first example illustrates risk assessment within decision-making processes relating to the siting of waste incinerators in the UK. This is site-specific risk assessment where discussion and debate is between proponent, decision authority and local communities. The development of the use of risk assessment in such siting debates has been driven by public rather than regulatory demand, questioning of scientific judgements, particularly over potential health risks from emissions, and concerns over the management of the risks.[26] The case example discussed here was the first quantitative risk assessment to be used in such a siting decision. Since then (1989–1991) there have been at least a further 17 similar cases.[27]

In the UK, risk assessments for new waste facilities are relevant not only to the planning permission, which will normally include a requirement for an Environmental Impact Assessment and publication of an Environmental Statement, but also to the subsequent required authorization or licence.[27] Problems arise because (i) the level of detail of the assessment may be different at the planning stage when the design of the facility may not be finalized, (ii) different regulatory bodies are involved, (iii) the risk management decision at the planning stage requires a balancing of risk *versus* other material considerations such as other environmental impacts, the issue of need and socio-economic impacts, and (iv) opportunities for public access and questioning are primarily at the planning stage, with the later stages reduced more to public notification. Not only

[23] T. Earle, and G. Cvetkovich, in *Risk Analysis*, ed. B. Zervos, Plenum Press, New York, 1993, p. 449.

[24] M. Schwarz, *Sci. Public Policy*, 1993, **20**, 381.

[25] B. Wynne, in *Social Theories of Risk*, ed. S. Krimsky and D. Golding, Praeger, Westport, CT, 1992, p. 275.

[26] J. Petts, in *Handbook of Environmental Risk Assessment*, ed. P. Calow, Blackwell Scientific, Oxford, 1997, p. 416.

[27] J. Petts and G. Eduljee, *Environmental Impact Assessment for Waste Treatment and Disposal Facilities*, Wiley, Chichester, 1994.

providing for a strong preventative regime in relation to environmental risks, the openness of the land-use planning system has provided for direct public questioning of proposals and proposed risk reduction measures, and for public discussion of social priorities and the values that people place on environmental quality. The planning system also provides for significant equity issues to be addressed, more specifically based on questions such as 'is this facility needed?' [*i.e.* a justification that it is the Best Practicable Environmental Option (BPEO)] and 'why here?'.

The application for a chemical waste incinerator at Seal Sands, Billingham, in north-east England was accompanied by a full, public Environmental Statement which included a quantitative risk assessment related to the emissions.[28] The application was refused by the local planning authority following considerable local debate and review of the assessment. The political decision was partly founded on a view that, given the evidence currently available and the disagreement over the environmental effects of incinerator emissions, the Council could not be satisfied that an incinerator would not endanger public health nor have a detrimental effect on the quality of life and the environment. The proponent appealed against the decision and a public inquiry was held. Two contemporary applications had also been submitted, and refused, for integrated treatment facilities to handle both sewage sludge and chemical wastes at Howdon, North Tyneside, and Portrack, Stockton-on-Tees. The inquiry dealt with the three appeals through the use of a rare inquiry commission to address the questions over 'need' and the BPEO, followed by the site-specific considerations.

The environmental statement for the Seal Sands facility addressed the atmospheric environment and human health impacts by an assessment largely following US risk assessment protocols. The assessment used emission dispersion modelling based on maximum concentrations of limit values for inorganics and conservative assumptions for organics. The quantitative risk from inhalation exposure was calculated with worst-case exposure assumptions. Non-carcinogenic risks were assessed by use of Occupational Exposure Levels adjusted by a safety value of 100 to account for differences in human sensitivity and residential exposure. Carcinogenic risks were assessed using the preferred US linear model with risks presented as the increased lifetime risk of developing cancer. Indirect exposure *via* the foodchain was calculated using the US Terrestrial Foodchain Model with worst-case assumptions related to food intake. The results were compared with acceptable daily intakes. The total incremental risk was calculated by the combined inhalation and ingestion routes and compared with other lifetime risks.

At the inquiry the assessment in the Environmental Statement was extended to deal with a number of additional concerns: specifically, quantification of risks from a further 24 compounds not covered by regulatory controls; an examination of the toxicology of 14 chemicals of possible concern; consideration of the epidemiological evidence from other incinerators; consideration of the risks to animals; consideration of the additive properties of the emitted chemicals; and characterization and assessment of potential interactions between neighbouring

[28] J. Petts and G. Eduljee, in *Environmental Impact Assessment for Waste Treatment and Disposal Facilities*, Wiley, Chichester, 1994, p. 413.

plumes and emissions from the proposed plant. The inquiry inspector summarized the health risk evidence as the 'most complex' submitted to the inquiry. There were questions over the emission rates used, based on emission criteria subsequently tightened by European standards; the degree of accuracy attached to the forecasts of maximum ground-level concentrations; the relationship between worst-case estimates and actual likely exposure; the influence of abnormal episodic releases; and the use of air quality standards related partly to practicability rather than health protection.[27] The inquiry, being a highly adversarial process, provided for representative input rather than direct public input. Despite this, there is evidence that public concerns as had first emerged at the planning application stage remained inherent in the debate.

The decision of the inspector was to grant permission to the Seal Sands application but not to the two other proposals. There were planning reasons why the latter were not granted, the inspector and his assessors being 'impressed by the (Seal Sands applicant's) concession that location away from houses was desirable on health grounds because if there is no exposure then there is no risk'.[29] The Seal Sands assessment gained credibility from its procedure compared to the assessments attached to the other two proposals. Although considerable uncertainties remained and these might be considered to weigh against the proposal, the Inspector considered that outstanding questions would not be sufficient on their own to warrant refusal and that control would be achieved through the Environmental Protection Act 1990.

Examination of this case and of subsequent incinerator applications[30,31] highlights the role of risk assessment in such decisions in promoting public accountability and confidence. There is some evidence that risk assessments in planning decisions support a view of the public as quality assurers in the risk management process.[32] The public focus tends not to be upon the actual risk estimate *per se* but upon the data inputs, assumptions and uncertainty management approaches used by the risk assessors. This not only challenges an expert focus on communicating risk estimates but also places the interface between the risk assessor and the public in the decision context under extreme pressure.

4 Risk Ranking and Site Prioritization in the Management of Contaminated Land

The second example relates to the use of risk assessment at the strategic level and, most particularly, to its use to prioritize limited resources so as to deal with the most significant risks. Risk assessment in this context is normally semi-quantitative or even qualitative. It has been developed in part, at least, as a means of justifying actions to a more distant public than was considered in the last case. It enables the intuitive expertise of experienced officers to be captured and passed on to more

[29] R. D. Donnison, *The Durham, Middlesbrough, Wallsend and Billingham Linked Inquiries into Incinerator Developments*, Department of the Environment, London, 1992, vol. 6.
[30] J. Petts, *Waste Manage. Res.*, 1992, **10**, 169.
[31] J. Petts, *Waste Manage. Res.*, 1994, **12**, 207.
[32] S. O. Funtowitz and J. E. Ravetz, *Environ. Toxicol. Chem.*, 1994, **13**, 1881.

inexperienced personnel in a standardized form, so potentially optimizing consistency in risk management decision making.

Contaminated land is a generic term encompassing a wide range of sites, pollution characteristics and health threats. Many contamination problems originate from the disposal of wastes, either deliberate or accidental. Contaminated land risk management has been significantly affected by the identification of 'problem sites'. Sites such as Love Canal in the US and Lekkerkerk in the Netherlands provide a number of risk management lessons: (i) they relate to known and deliberate disposal of wastes, the consequences of which were neither fully recognized nor understood at the time; (ii) they represented neither examples of necessarily 'bad' practice nor illegal practice; (iii) risk management such as existed was reliant on ongoing institutional control which was not forthcoming; and (iv) public awareness of the problems and ensuing alarm prompted regulatory intervention.[33] The direct outcome of Love Canal was the Comprehensive Environmental Response, Compensation and Liability Act 1980 (CERCLA): the so-called 'Superfund' legislation. CERCLA authorized the federal government to respond directly to releases or threatened releases of hazardous substances that may endanger human health, welfare or the environment.

Subsequent federal action led to the formation of a national priority list (NPL), which consisted of a league table of approximately 1200 contaminated sites. A screening methodology known as the Hazard Ranking System (HRS) has played a central role in classifying these sites, shaping the definition of the contaminated land problem. It was designed as a means for applying uniform technical judgement to a range of sites which would indicate the level of hazard relative to other sites.

Hazard and risk assessment experience since the 1950s highlights the importance of isolating a closed system which defines the problem. The HRS does this by identifying nearly 100 individual factors considered by experienced staff to be indicative of the risk posed by contaminated land. The lack of adequate information about the problem makes this initial step essential but does require explicit recognition of the limitation of the assessment model in this respect. A keen awareness of a model's limitations will improve the chances of it being used effectively. However, experience shows that such acknowledgement is not always apparent.

Many forms of structured value systems have been used to evaluate particular characteristics of contaminated sites. Such systems typically consist of a framework of individual factors or characteristics that are logically grouped. Table 1 illustrates a typical hierarchical system. The HRS structure is based on three key hazard modes. The migration potential hazard mode is applicable to longer term remediation whilst the fire and explosion and direct contact hazard modes are used to determine the need for emergency action. Each is divided into categories which are further subdivided into factors. Scores are attributed to each of the factors according to prescribed guidelines. These are combined using a simple algorithm, usually using some form of weighting system, to produce a

[33] J. Petts, T. C. Cairney and M. Smith, *Risk-Based Contaminated Land Investigation and Assessment*, Wiley, Chichester, 1997, p. 15.

Table 1 The hierarchical structure of structured value systems

Level in hierarchy	Description	Example from a typical assessment model for contaminated land
Top-level overall score	Ultimately an *OVERALL SITE SCORE* can be generated through the combination of components that may or may not have been weighted	Standardized numerical value often expressed as a percentage or in relation to the worst possible case
High-level components	Groups of categories that have been scored and weighted are termed *COMPONENTS*	Characteristics of the waste Engineering containment Migration potential Potential targets
Mid-level categories	*CATEGORIES* are groups of individual factors that have been scored and weighted	The migration potential component might consist of: groundwater pathway surface water pathway air pathway direct contact
Bottom level factors	Baseline *FACTORS* represent the lowest level of a model; these are the individual characteristics that are initially scored	The groundwater pathway category might consist of: local geology distance to aquifer sensitivity of the aquifer

single score, in this case representing the overall score which may be considered indicative of the risk posed by that site.

The crucial point is that the structure and values are generated subjectively by experts, often by brainstorming exercises designed to elicit all of the features of the hazard generating system that are indicative or contribute in some way to the risk. Having established the structure, the numerical values are then devised using prescribed scoring and weighting systems again developed by expert elicitation. Each of these processes is open to a number of potential distortions ranging from inevitable, individual, idiosyncratic traits and random noise to bargaining behaviour and even deliberate manipulation arising from organizational biases.[34] Thus, the processes of modelling complex systems using mass aggregation and over-simplification may not serve to clarify issues but to conceal them. For example, a bias towards the groundwater component of the HRS has been argued to explain why fewer contaminated sites have been identified in large densely populated urban areas which have a dependence on surface water supplies.[35] This has arisen as a potential problem because these urban centres have a higher than average minority population and there is particular concern at present about the social equity issues surrounding the distribution of risk and benefits.

The way in which factors and weights are developed and revised is crucial to the model. As revisions are made, so site rankings will alter, bringing new sites to

[34] J. Forester, *J. Am. Planning Assoc.*, 1982, **48**, 67.
[35] R. Zimmerman, *Risk Anal.*, 1993, **13**, 649.

Table 2 The original HRS scoring prescription for hazardous waste quantity

Tons/cubic yards	No. of drums	Assigned value
0	0	0
1–10	1–40	1
11–62	41–250	2
63–125	251–500	3
126–250	501–1000	4
251–625	1001–2500	5
626–1250	2501–5000	6
1251–2500	5001–10 000	7
> 2500	> 10 000	8

the fore and relegating others. The following example of the development of just one of the HRS factors illustrates the way in which revisions have occurred.

Incorporating Quantitative Data: Managing Waste Quantity

Despite the importance of understanding something about waste quantity at a contaminated site, initially the waste quantity factor in the HRS was relatively insensitive. According to an unpublished sensitivity analysis of HRS factors, in the ground and surface water pathways the waste quantity factor was the least sensitive of all the HRS factors.

Our understanding, based on interviews with Environmental Protection Agency Site Assessment Branch officials, indicates that this insensitivity resulted from the inherent complexity of the factors within the HRS: an unintentional by-product of design arising from the mass of competing factors, rather than from some deliberate policy.

The original waste quantity scoring prescription[36] is outlined in Table 2 and typifies the nature of the entire HRS scoring system. In this version, where waste quantity is unknown the assigned default allocates one point and the maximum score is eight points. No account is given as to whether or not the presence of 10 000 drums is eight times worse than for a single drum, nor is the basis of this assumption explained.

Revising Waste Quantity

Of the 1286 sites on the NPL in 1986, some 269 (roughly 20%) showed no information for waste quantity and thus had been scored with the default value. As waste quantity was considered to be intuitively important, yet so numerically insensitive, the United States Environmental Protection Agency (US EPA) was keen to revise its status. Three issue analyses were commissioned, concerning the use of the default value for sites where waste quantity data were unavailable,[37] the potential for using hazardous constituent data instead of waste quantities[38] and

---

[36] US EPA, *Fed. Regis.*, 1982, **47**, 31 180.

[37] L. M. Kushner, *Hazard Ranking System Issue Analysis: Sites With Unknown Waste Quantity*, Report MTR-86W83, Mitre Corporation, McLean, VA, 1986.

[38] A. Wusterbarth, *Hazard Ranking System Issue Analysis: Relationship Between Waste Quantity and Hazard Constituent Quantity*, Report MTR-86W00141, Mitre Corporation, McLean, VA, 1986.

Table 3 The revised HRS scoring prescription for hazardous waste quantity

Tier	Measure	Units	Equation for assigning value
A	Hazardous constituent quantity (C)	lb	C
B	Hazardous waste stream quantity (W)	lb	$W/5000$
C	Volume (V)		
	Landfill	yd^3	$V/2500$
	Surface impoundment	yd^3	$V/2.5$
	Surface impoundment (buried or backfilled)	yd^3	$V/2.5$
	Drums	gallon	$V/500$
	Tanks and containers other than drums	yd^3	$V/2.5$
	Contaminated soil	yd^3	$V/2500$
	Pile	yd^3	$V/2.5$
	Other	yd^3	$V/2.5$
D	Area (A)		
	Landfill	ft^2	$A/3400$
	Surface impoundment	ft^2	$A/13$
	Surface impoundment (buried or backfilled)	ft^2	$A/13$
	Land treatment	ft^2	$A/270$
	Pile	ft^2	$A/13$
	Contaminated soil	ft^2	$A/34000$

the potential for using hazardous substance concentrations.[39] An investigation was undertaken to determine the feasibility of deriving hazardous substance concentrations for different sources of waste at contaminated sites such as drummed wastes, lagoons, mine tailings and landfill sites. Given the paucity of waste data this was an ambitious exercise, and not surprisingly the report could only reach the tentative conclusion that drummed waste was likely to contain higher concentrations of hazardous materials than wastes stored in open lagoons which were likely to have been diluted by rainwater.

One solution to the inherently subjective nature of the required judgements was to adopt a tiered approach to managing data quality such as that objective data were assigned proportionately greater weight. The proposed revision for waste quantity adopted a scoring basis which utilized actual amounts of materials found at the site, though judgement was still necessary to estimate the volumes of waste materials. The proposed revision of the waste quantity factor included a maximum value of 1000 lbs. The final rule increased this ceiling to 1 000 000 lbs. As linear scales for hazardous waste quantity, toxicity and other factors are multiplied in the HRS algorithm, the US EPA found it necessary to adopt a scaling transformation approach. As Table 3 shows, it became necessary to divide some of the larger scores by scaling factors in order to ensure that the particular scores generated were not overly influential.[40] Whilst this meant that the waste characteristics category did not have a disproportionate impact on the likelihood of release, or targets categories, it inevitably weakened the site data.

[39] Mitre Corporation, *Hazard Ranking System Issue Analysis: Consideration of Contaminant Concentration*, Report MTR-86W40, Mitre Corporation, McLean, VA, 1987.

[40] US EPA, *Fed. Regis.*, 1990, **55**, 51 591.

HRS is a complex semi-quantitative hazard ranking tool, not designed to produce site-specific risk assessments but to prioritize management responses which in most cases results in more detailed risk assessment. The US system of model development is relatively open to public comment, with statutory review periods which allow interested parties to influence the risk assessment process, albeit often through costly law suits. Study of the development of HRS since the early 1980s reveals key questions relevant to the design of all such systems, not least who possesses relevant expertise, how is this elicited in an objective way such that biases are identified, how is data uncertainty managed and made explicit, and perhaps most importantly, who decides on the relative balance of inputs and the criteria for evaluating risk? The revisions to HRS discussed above illustrate a desire to provide more flexibility in the system without imposing any significant extra costs on data collection. This desire for cost efficiency was one of the first signs that economic concerns were impacting on HRS. Cleaning up abandoned hazardous waste sites has become one of the most costly undertakings of US environmental policy, with projected costs reaching hundreds of billions of dollars into the new millennium.[41] Risk assessment approaches have to meet social and economic priorities. These must be made explicit in development of the approach.

5 The Development of Operator and Pollution Risk Appraisal (OPRA)

The third example is also at the tier of strategic regulation, though this time in the context of a UK regulatory agency. It addresses the role of risk assessment in the focusing of resources and the consistent application of regulation.

In the US, the majority of efforts to provide environmental protection have been based on formal legislation, where courts have the power to review and enforce standards. The uniformity of regulation is regarded as part of the notion of procedural fairness. By contrast, the UK approach is founded more closely on the ideal of technical efficiency enforced informally with administrative discretion.[42] Less emphasis has been placed on legally enforceable controls, in part through a strong desire not to impose too heavy a financial burden on industry. There has long been a trade-off between the desire for technical and economic efficiency, and political responsiveness to public anxieties and demands. In comparison to the US, the UK might be said to lean more to the former than the latter, with less structured mechanisms for public comment and more consensual relationships between industry and regulators.

Increasing political promotion of voluntary action to support regulatory control might suggest the relegation of inspection to a position of secondary importance. There is growing evidence from risk management controversies that the public lack confidence in the regulator. This lack of trust impacts directly on public perception of risk and is sometimes converted into considerable opposition

---

[41] A. Wildavsky, *But is it True? A Citizens Guide to Environmental Health and Safety Issues*, Harvard University Press, Cambridge, MA, 1995.

[42] T. O'Riordan, A. Weale and L. Kramme, *Controlling Pollution in the Round. Change and Choice in Environmental Regulation in Britain and Germany*, Anglo-German Foundation, London, 1991.

to proposals by industry to develop hazardous facilities.[27,43] Thus, it is crucial that the regulator is able to inspire public confidence by adopting an effective inspection routine which is seen to keep industry in check with regard to their environmental performance.

OPRA, developed by Her Majesty's Inspectorate of Pollution (HMIP) and subsequently adopted by the Environment Agency (EA) (see also Section 3 of the Halfacree article in this volume), is intended to permit officers to assess industrial performance and set appropriate inspection frequencies through consideration of two key modes: Operator Performance Appraisal (OPA) and Pollution Hazard Appraisal (PHA). As with the ranking models described earlier, OPRA relies on the identification of indicative factors which are subjectively scored and weighted. The consultation document presented the prescriptive guidelines for scoring based on a 1–5 scale. The inspector evaluates the particular site and situation on the basis of two endpoints, a best and worst case, which are described in the accompanying guidelines to the model. However, the consultation document fails to make clear the precise nature of the weights applied to the scores, preferring to acknowledge that the weights 'may be modified following a period of usage and in the light of experience'.[44] This is despite the recognition that OPRA will 'work best if there is transparency between Inspectors and Operators, and the system is fully understood by all parties'.[45]

The translation of assessment scores into inspection frequencies is also problematic. The consultation draft presented an indicative banding of operator performance and pollution hazard which provides for site scores to be plotted on a graph and placed into high, average and low groups. How these bands are derived, for instance whether they are based upon the significance of potential environmental risk and the regulator's expected performance criteria or more reflective of manageable inspection frequencies, will have a significant impact on public acceptability. This requires such judgements to be made explicit and justified.

OPRA is a potentially important tool for the regulator as a means of prioritizing resources. However, OPRA has wider implications than budgetary management. The effectiveness of inspection is now closely linked to the public image of the regulator. OPRA goes some of the way to opening up the black box but it remains to be seen whether the model will stand up to the rigorous external scrutiny it could eventually receive.

6 Using Risk Assessment as a Means to Derive National Policy Objectives

OPRA and its associated tool for assessing the BPEO[46] reveal all of the assessment difficulties which face a regulatory regime based in a policy which

43 V. T. Covello, in *Risk Perception and Communication*, DoE Report No. DoE/HMIP/RR/95.011, Her Majesty's Inspectorate of Pollution, London, 1995, p. J-1.

44 Her Majesty's Inspectorate of Pollution, *Operator and Pollution Risk Appraisal*, HMIP, London, 1995, p. 4, para. 10.

45 Her Majesty's Inspectorate of Pollution, *Operator and Pollution Risk Appraisal*, HMIP, London, 1995, p. 4, para. 11.

46 Environment Agency, *Best Practicable Environment Option Assessments for Integrated Pollution Control*, HMSO, London, 1997, vols. 1 and 2.

focuses on integrated pollution prevention and control. In the UK, all government departments have to justify new regulations with a risk/cost assessment. This places not only significant pressures on the scientific input to the assessment in terms of relative environmental impacts, but also the need for greater public accountability in terms of risk/cost trade-offs. The UK Royal Commission on Environmental Pollution is completing a study on environmental standards which has included taking evidence on the need, potential and nature of stakeholder (including public) involvement in the development of standards. Comparative risk assessment and risk/cost assessment in policy making also suggests a potential tension between national policy and the acceptability of risk at the local level. Two areas of policy development highlight the difficulties for risk assessment at this higher tier: contaminated land clean-up criteria and emission limits for dioxins.

At the time of writing, new risk-based guideline values for the clean-up of contaminated sites are expected to be introduced in the UK. The criteria have been developed following criticism[47] of the restricted nature of the Interdepartmental Committee on the Redevelopment of Contaminated Land's (ICRCL) trigger values and the lack of transparency as to their basis. The new guideline values are derived using the Contaminated Land Exposure Assessment Model (CLEA).[48] CLEA is a probabilistic risk assessment model which appears to be the only one currently developed for deriving national contaminated land criteria.[49] Common to developments in the US, the increasing move to probabilistic methods as opposed to deterministic point estimates of risk reflects concern to account more objectively for uncertainty, although it is computationally more intensive and there is relatively little experience of its use in public decision fora (except in radioactive waste facility siting).

Adoption of CLEA into national guidelines for regulatory purposes has required decisions on a number of key questions, such as: (i) what percentage of the population should be protected (*e.g.* 95%, 90%), (ii) what proportion of a tolerable daily intake should come from a contaminated site compared to that from other potential exposures, and (iii) what is an 'acceptable' or 'negligible' additional cancer risk from a contaminated site, given that there will be background risks from other sources?[29] To answer these questions requires high-level policy decisions that should be equitable across media and fair to all parties. It is hoped that the justification for the decisions will be apparent in the guidance to be published with the guidelines.

Implementation of the guidelines at the local level will require local cost/risk trade-offs. It is not apparent that a level of soil contamination deemed to be acceptable nationally is necessarily acceptable at a specific site, or *vice versa*. Indeed, the development of risk assessment for environmental risk management in the UK has been recognized to provide a non-prescriptive approach to regulation and control which allows decisions to be made appropriate to the

---

[47] House of Commons Select Committee on the Environment, *Contaminated Land*, First Report, HMSO, London, 1990, vols. 1–3.

[48] C.C. Ferguson and J.M. Denner, *Land Contamin. Reclam.*, 1994, **2**, 117.

[49] C.C. Ferguson, *Land Contamin. Reclam.*, 1996, **4**, 159.

environmental setting.[50] While the guideline values will play an essential screening role in the risk assessment of specific sites, the site-specific acceptability of a risk will be dependent upon the characteristics of the targets, the strength of the source–pathway–target link and background exposure. The much welcomed drive to more objective and consistent assessments will still place pressures upon risk assessors' skills and upon the transparency of decision making.

International debate over dioxins has highlighted the problems of known scientific uncertainty in the development of appropriate controls, pressures to adopt the precautionary principle as a result of this uncertainty, and considerable scientific/expert disagreement, with consequent impact on public perceptions. The European emissions standard of $0.1 \, \text{ng m}^{-3}$ is based on progressive Best Available Technology. There are technical difficulties in monitoring for dioxins at such low concentrations (not least in obtaining a representative stack gas sample) and if there is pressure to translate the standard into other directives (for example, relating to municipal waste incineration) this could result in increased costs and encourage a less environmentally acceptable method of disposal. A risk assessment of dioxin releases from municipal waste incineration processes for the UK government[51] concluded that at $1 \, \text{ng m}^{-3}$ (the current UK standard) there would be no significant risk to exposed individuals irrespective of the location and size of the plant, or the characteristics of the exposed population. The assessment was based upon a tolerable daily intake of 10 pg I-TEQ (kg body weight)$^{-1}$ day^{-1} endorsed by the UK government, but subject to challenge arising from the US EPA assessments (I-TEQ, International Toxic Equivalents).

Whether the focus on incineration compared to other sources of (toxic equivalent) dioxins is appropriate in terms of comparative risks is open to debate, although there is little doubt that incineration is easy to target in the regulatory context and controls do not appear to have direct social impacts (rising waste disposal costs being largely hidden). Risk assessment for the siting of incinerators (as discussed earlier) provides a site-specific check on the protection provided by national standards. In the public domain an argument that operation to a national standard provides proof of acceptable risk at a specific site is not acceptable.

Risk assessment at the policy level is fundamentally divorced from direct public influence, although peer review is important. Discussions of the importance of Strategic Environmental Assessment (SEA) for policies and plans has revealed a tension over confidentiality at this level, which belies the general belief that public participation in SEA is fundamental to the process.[52] The US EPA has used regulatory negotiation to derive acceptable risk/benefit trade-offs: for example, over recycling of lead acid batteries, emission standards for architectural and industrial coatings and wood stoves emissions.[46] However, as a mechanism for the lay public to participate in making environmental risk decisions the process is still deficient. Negotiation may involve bias in favour of the existing distribution

[50] Department of the Environment, *Guide to Risk Assessment and Risk Management for Environmental Protection*, HMSO, London, 1995.

[51] Her Majesty's Inspectorate of Pollution, *Risk Assessment of Dioxin Releases from Municipal Waste Incineration Processes*, Report No. HMIP/CPR2/41/1/181, HMIP, London, 1996.

[52] B. Sadler *Environmental Assessment in a Changing World: Evaluating Practice to Improve Performance*, Canadian Environmental Assessment Agency, Ottawa, 1996, p. 139.

of power and resources in society and interests that are not recognized at the national level lack access to the negotiations.[53] The extent to which risk assessment can be open to more direct public input and involvement is challenged by ingrained institutional procedures as well as by expert reluctance to recognize or accept the potential benefits of public input.

7 Conclusions

The discussion of risk assessment application at different tiers of decision making has highlighted a number of key points. First, although scientific and technical uncertainty is accepted as a significant element of risk assessment, objective means are available to manage this, not least through sensitivity analysis and probabilistic approaches. However, expert willingness to accept the impact upon uncertainty of professional bias, lack of skills in risk assessment, bargaining behaviour, organizational priorities, *etc.*, has not always been as strong, and yet in the public domain these uncertainties have become increasingly visible.

Linked to this, over time, risk assessment has become more of a procedural than a substantive concern. The 'due process' which is followed in conducting risk assessments is now as influential as, if not more than, the results themselves. This is particularly the case in the US and becoming increasingly so in the UK. However, in the UK the consensual rather than adversarial culture between the regulator and industry means that risk assessment techniques, particularly those used for prioritization, still remain largely a 'black-box' to the lay public.

Comparative risk assessment is now driving the national and international agenda, although whether and how to make risk/benefit trade-offs between essentially 'apples and oranges' problems could remain the primary challenge of the next decade. It has been argued that people compare apples and oranges not using quantitative factors such as the relative concentration of specific vitamins but through more subjective factors such as flavour and texture.[54] Though this might seem a simplistic analogy, it holds some merit and points to a future which acknowledges the rightful place of subjective, qualitative factors within the overall assessment process. The choice is not between a wholly objective or subjective assessment but between one which explicitly recognizes the inherent uncertainties and subjectivities while still trying to foster sensible, sophisticated and informed debate and one which does not.

The number of residents' groups which are now challenging successfully the scientific community on their own grounds is increasing rapidly.[55] Citizen knowledge has been mobilizing, particularly in the US, with subjects such as popular epidemiology gaining credence with public administrators and the courts.[56] This has led to local groups being funded to prepare their own cases

[53] D. Fiorino, in *Fairness and Competence in Citizen Participation; Evaluating Models for Environmental Discourse*, ed. O. Renn, T. Webler and P. Wiedemann, Kluwer Academic, Dordrecht, 1995, p. 223.

[54] A.M. Finkel, in *Worst Things First? The Debate over Risk-Based National Environmental Priorities*, ed. A. M. Finkel and D. Golding, Resources for the Future, Washington, DC, 1994, p. 337.

[55] M. K. Heiman, *J. Planning Educ. Res.*, 1997, **16**, 137.

[56] P. Brown and E. J. Mikkelsen, *No Safe Place. Toxic Waste, Leukemia, and Community Action*, University of California Press, Berkeley, CA, 1990.

using their own consultants and learning how to design and implement their own monitoring programmes. Recent challenges by local action groups in the UK have had similar successes, albeit on a much smaller scale, in halting development of proposed waste facilities and altering the licence arrangements of existing plant through challenges to the scientific procedure of risk assessments. The next decade will see an increasingly willing and able public which can readily access risk assessments from sources worldwide. The focus of expertise no longer lies solely with the scientist or technical expert.

Regaining some control over the scientific procedure may not in the end reduce conflict over controversial proposals, but will serve to highlight the uncertainties and subjectivities in risk assessments which, in turn, should improve the quality of debate and the final decision. The US EPA's Technical Assistance Programme provides up to $50 000 to community organizations for monitoring site remediation.[57] There are other potential benefits from this opening up of the scientific procedure, which include: a greater level of understanding of the basis of risk assessments within groups that might otherwise never have considered them; a greater sense of control over the decision-making process from which local people may have felt disenfranchized; and a greater sympathy for the decision maker through an increased understanding of the complexity of problems and choices.

[57] R. R. Kuehn, *Univ. Illinois Law Rev.*, 1996, **1**, 163.

The Use of Risk Assessment in Government

JIM McQUAID AND JEAN-MARIE LE GUEN

1 Risk and Government

Greater prosperity and improved standards of living have been accompanied, perhaps not surprisingly, by greater public demand for healthier workplaces, cleaner environment, better housing, safer food and commercial products. More generally, there is a hankering for a society free of involuntary risks, underpinned by a belief that the State can do much to insulate people from such risks, or, in the event that they are realized, mitigate their consequences. The result is that there is an increasing frequency of demands for explanations of how it is intended to address risk issues, both in general and in particular circumstances, in a way that is acceptable to the public. Indeed, so much is this so nowadays that issues pertaining to the distribution of risks receive as much attention in the media as those pertaining to the distribution of wealth. This, together with the need to adapt to global changes characterized by the emergence of new technologies, improved communications, intensified economic competition and increased co-operation between countries, has led to pressures for new approaches to the regulation of risk. What is emerging is a clear need for the development and implementation of consistent and workable policies and practices to take account of all the above factors.

In the UK, the Inter-Departmental Liaison Group on Risk Assessment (ILGRA) is organized by the Health and Safety Executive on behalf of all Government Departments in order to provide a mechanism for the development of consistent and coherent policies and practices on risk assessment. It is involved in promoting new approaches, fostering research on topics of cross-Government relevance and maintaining communication links between those actively engaged in policy development on risk assessment in different Departments. In this article, the context for the work of ILGRA is surveyed and the various frameworks within which it seeks to improve the quality of risk regulation are described.

2 Problems

There are many—some seemingly intractable—problems that Government must tackle in the area of risk regulation. Four examples of these are discussed here.

Adoption of an Approach on Risk Compatible with the Trend for Less Government

Although the public want to be shielded from risks and compensated when something goes wrong and they suffer some detriment, at the same time they resent regulations that they regard as needlessly restricting fundamental rights or which introduce 'red-tape' and burdensome administrative formalities. Certain initiatives by the media to expose and denounce instances of red-tape have promoted a perception that 'regulatory inflation' may be endemic, with effects variously described as curtailing unnecessarily the freedom of individuals, creating a dependency culture, slowing down economic growth, discouraging job creation and more generally introducing rigidities which have anti-competitive effects.[1]

This perception, though, has not resulted in a situation where regulation *per se* is believed to be undesirable. The general awareness remains that regulations do bring benefits because they do reduce risks, introduce a level playing field and give a general assurance that the State has systems in place to ensure that industry can exploit technological advances and breakthroughs without letting the public face unacceptable risks. There is thus the paradox that calls for more regulation are often made by the public, the media, pressure groups, industry, trade unions, Parliament and Governmental institutions as a response to individual issues considered in isolation.

There are areas such as environmental quality where the Government is responding to such calls by introducing more regulation. However, the past practice of responding slavishly to pressures for more regulation is not in keeping with present-day calls for less government. Alternatives to regulation need to be considered. Where regulation is still felt to be necessary, it is now accepted that the regulatory requirement should be framed so as to achieve an effect proportionate to the risk. Only in this way can a workable balance be found between conflicting pressures for more regulation on the one hand and less Government intervention on the other.

It is Getting Harder to Characterize Risks

There is a long history of Government regulating to minimize risks. The first ones to be regulated were the most obvious and often required little scientific insight for identifying the problem and possible solutions. For example, it was not difficult to realize that regulating to control airborne dust would reduce the risk of silicosis in miners and that regulating to control emission of smoke from factory and domestic chimneys would reduce the incidence of lung disease. In short, dramatic progress towards tackling such problems could be made without taxing existing scientific knowledge or the state of available technology.

However, as the most obvious risks have been tackled, new and less visible hazards have emerged and gained prominence. Typical examples include those arising from new technologies such as nuclear power and biotechnology, and processes emitting gases which contribute to global warming and ozone

---

[1] J. Adams, *Risk*, UCL Press, London, 1996, pp. 205–208.

depletion. One frequent characteristic of these new hazards is that it can be very difficult to define precisely the risks they may give rise to, even when scientific knowledge is pushed to the limit. The processes that may give rise to risks are only partially understood, with the result that regulatory decisions must frequently be based on limited data and considerable scientific and technological uncertainties.

Increased Complexity in the Regulation of Risk

The regulation of risk is becoming increasingly complex. For example, the regulatory framework for occupational health and safety in the UK in the last two decades has been built upon the enabling legislation of the Health and Safety at Work Act 1974, which imposes general duties on those who create risks to ensure that risks are prevented or adequately controlled. Since then, regulations have been introduced extending these duties further by requiring people, such as employers and self-employed persons, to assess risks and to base their control measures on the results of the assessments. The trend in recent years has been to amplify these duties still further, for circumstances where high hazards are involved, to require the production of safety cases where duty holders have to write down and submit to the regulator the measures they intend to introduce to meet their legal obligations.

Furthermore, whereas there was a time when agreement on the action necessary could be reached on the basis of the degree of risk posed by a particular hazard as assessed by applying theories from natural sciences, engineering, logic and mathematics, this is no longer the case. This approach is no longer sufficient to counter the growing demand that regulation of risks should take account of the quality of risk as distinct from objective assessment of the quantum of risk. It has become a matter of course to request, for example, that undesirable consequences should include consideration of matters such as distributional or economic equity or ethical considerations,[2–4] or whether for environmental risks it is morally right to adopt policies without considering the effects on natural phenomena like the survival of species and the maintenance of ecosystems. In short, the concept of risk has evolved to include matters that do not readily lend themselves to scientific analysis.

Indeed, there is a well-established school of thought[5–8] that it may be futile nowadays to believe that there is a quantifiable physical reality that can be derived which people will agree represents the risk from a hazard. Proponents argue that the concept of risk is deeply ingrained in human nature and has helped

[2] L. R. Beach, *Image Theory: Decision Making in Personal and Organizational Contexts*, Wiley, New York, 1992.

[3] P. M. Sandman, N. D. Weinstein and M. L. Klotz, *J. Commun.*, 1987, **37**, 93.

[4] E. Vaughan and M. Seifert, *J. Social Issues*, 1992, **48** (4), 119.

[5] M. S. Douglas and A. Wildavsky, *Risk and Culture*, University of California Press, Berkeley, CA, 1982.

[6] S. O. Funtowicz and J. R. Ravetz, in *Social Theories of Risk*, ed. S. Krimsky and D. Golding, Praeger, Westport, CT, 1992, pp. 251–274.

[7] N. C. Pidgeon, C. Hood, D. Jones, B. Turner and R. Gibson, *Risk: Analysis, Perception and Management*, The Royal Society, London 1992, pp. 89–134.

[8] B. Wynne, in *Social Theories of Risk*, ed. S. Krimsky and D. Golding, Praeger, Westport, CT, 1992, pp. 275–300.

mankind in its evolution to cope with the dangers and uncertainty of life. However, this concept of risk is shaped by human minds and cultures and though it may include the prospect of physical harm, it may include other factors as well. The logical conclusion from that theory is that it is judgement and values that determine which factors are defined in terms of risk and actually subject to measurement.[9]

This idea that risk is multi-dimensional is now well established.[10] People are more averse to certain risks than others. Important factors affecting such perception include how well the processes giving rise to the hazard are understood, how equitably the risk is distributed, how well individuals can control their exposure to the risk and whether it is assumed voluntarily.[11] This helps to explain why, for many new hazards, high quality risk analysis by leaders in the field often lacks authority. Even using best available science, such risk analysis cannot be undertaken without making a number of assumptions. Parties who do not share the judgement values implicit in those assumptions may well see the outcome of the exercise as invalid, illegitimate or even not pertinent to the problem—as exemplified by the controversy surrounding the proposal to sink the Brent Spar oil platform in the middle of the ocean.

Changes in the Regulatory Environment

The regulatory environment now has to cope with the increasing trend in industry and elsewhere to outsource risks, leading to changes in patterns of employment. There have been dramatic increases in self-employment and home-working, and small and medium size firms are a major force in creating jobs. These changes are making it more difficult to adopt the traditional approach of imposing legal duties on those who can take steps to control risks because it is often not easy to determine who are in such a position. In short, methods for regulating risks that worked well when large firms provided the bulk of employment seem no longer appropriate.

In addition, the regulation of risk is increasingly being undertaken at European or international level in the wake of the creation of new global markets and new technologies. These developments again beg the question as to whether long-standing regulatory traditions are the best way forward. For some of the new risks, like those arising as a result of global warming, ozone depletion and the release of genetically modified organisms, action will clearly have to be taken at international level to have any effect. Moreover, in other areas the technology is moving so fast that *de facto* international standards or practices are evolving all the time, *e.g.* in ensuring the safe use of computerized systems for controlling plant and machinery. Many countries are calling for such technologies to be regulated at international level as the only effective way to prescribe appropriate standards.

The move towards the internationalization of regulation not only requires innovative forms of regulatory co-operation but must take into account hosts of

[9] Ref. 1, pp. 7–27.

[10] P. Slovic, *Science*, 1987, **236**, 280.

[11] B. Fischoff, P. Slovic, S. Lichtenstein, S. Read and B. Combs, *Policy Sci.*, 1978, **9**, 127.

other factors such as agreements for regulatory harmonization, mutual recognition of standards and removal of barriers to trade.

3   Solutions

The approach to solving the above problems rests, amongst other things, on:

- Adopting quality regulation
- Strengthening liaison arrangements between Departments
- Improving systems for communicating information on risks to the public
- Strengthening the links with non-governmental organizations and academic institutions

These are examined in further detail below.

4   Adopting Quality Regulation

The regulation of risk used to be regarded as primarily a technical activity which by and large could be carried out internally on the basis of a department's or regulator's objective analysis of the degree of risk. Now that the need to take account of political and social factors has come more to the fore, there is an increased emphasis on assuring the quality of regulation. Such an approach can go a long way in reassuring the public that there are adequate safeguards for protecting them from risks, while getting the balance right between individual and collective rights and the degree of Government intervention.[12] In practice, quality regulation should ensure that stakeholders affected by the risk can actively participate in the consideration of regulatory proposals and that enforcement of the law should:

- Require that actions taken are proportionate to the risks
- Be targeted on the most serious risks or on hazards that are less well controlled
- Be consistent by ensuring a similar approach in similar circumstances to achieve similar ends
- Be transparent for all to see how decisions were arrived at and what they mean for them

The achievement of these objectives requires the adoption of specific procedures as described below in five stages. The boundaries between these stages are not clear cut. Valuable new perspectives may arise as one progresses through each stage. In short, the procedures are governed by the principle of iteration, as will be apparent as each stage is examined in more detail.

Stage 1: Defining the Problem—Assessing Risks

The route to good regulation for health and safety lies in being able to identify with confidence what is the risk problem to be solved, the nature and magnitude

---

[12] Cabinet Office, *The Regulatory Appraisal Guide*, Better Regulation Unit, Cabinet Office, London, 1997.

of the risk, who is likely to be affected by it, and how those affected and society at large perceive the problem. Thus there will be a need:

- To undertake a risk assessment or examine those situations where harm or damage could arise, the likelihood of this occurring and what the consequences would be. In practice, this may not be easy. It is unlikely that the risk assessment can be completed without making a number of assumptions for covering gaps in data and/or knowledge of the underlying science
- To ensure that those affected by the risk are convinced that the risk assessment is valid. As such, the important aspects of the risk assessment are not only the number or judgement arrived at, but also the evidence used in arriving at it and the reasoning it encompasses. In practice, this means being open about the reasons for making assumptions and how they were reached and giving an indication of their robustness

Stage 2: Examining the Options Available for Regulating and their Merits

This stage can be broken into two steps: identifying the options and assessing their merits. The first step calls for imagination and creativity to identify the span of actions available. This can range from doing nothing to introducing measures (whether non-regulatory or regulatory) to get rid of the cause of the problem altogether or to reduce it to one which people are prepared to live with.

For example, the following options (many of which are not mutually exclusive) have been suggested for preventing pollution by a particular substance: banning the use of the substance altogether; using technology for preventing the substance being released to the environment; introducing licensing regimes to limit the number of people using the substance while ensuring that they use best practice to prevent accidental release of the substance; taxing industry on the basis of the pollutants they release; educating or informing the public on the steps they can take to prevent pollution, *e.g.* on the safe disposal of nickel–cadmium batteries.

Cost–Benefit Analysis. It will also prove useful at this stage to undertake a cost–benefit analysis (CBA) to obtain data on the balance between the benefits of each option against the costs incurred. CBA aims to express all relevant costs and benefits in a common currency, usually money. This, in principle, requires the explicit valuation of the benefit of reducing the risk. However, such a valuation may not always be possible or practicable; in these and indeed many other situations, common sense judgement may have to be applied.

Where an option is about avoiding fatalities, the benefit of the option is a reduction in the risk of death. For example, when conducting CBAs in support of policy proposals, the Health and Safety Executive (HSE) currently uses as a baseline the proposition that a reduction in the individual probability of death of ten in a million per year is worth about £8. This figure derives from the value used by the Department of Transport for the appraisal of new road schemes. However, the HSE would regard higher values as being appropriate for certain risks, *e.g.* those which give rise to high aversion.

There will, of course, be many options where potential benefits are not concerned with a reduction in the risk of death; for example, avoiding deafness or dermatitis or a major injury. Very often in these cases, monetary values can be placed on a reduction of the risk by comparing how society views detriments such as a major injury in relation to death. In addition, there may be non-monetary benefits of a regulatory option, such as improvement in the sense of well-being or security.

Expected costs for an option may be non-monetary as well as monetary. Typical examples of monetary costs include those associated with the development and application of technology, training, clean-up, *etc.* Non-monetary costs include loss of things that people value, such as convenience or a reduction in flexibility, and choice for consumers and business, for example if a product or process is banned.

Potential Adverse Consequences.   It is also necessary to consider at this stage whether an option may cause any adverse consequences. Frequently, adopting an option for reducing one particular risk of concern may create or increase another type of risk, *e.g.* banning a particular solvent may increase the use of a more hazardous one because there is no other substitute; reducing concentration of airborne substances in the work place by exhaust ventilation increases risk in the community outside. For each option having adverse consequences it will be necessary to examine the trade-off between the reduction in the existing risk and the one being created or increased.

Constraints.   It will also be necessary to identify any constraints attached to a particular option, *e.g.* the option may require the development of substitute processes or substances; there might be legal constraints on its adoption, such as incompatibility with existing national or European Community legislation or international norms or agreements; or its adoption may give rise to ethical issues because some people may consider it as legitimizing exposure to a risk that they believe, on moral grounds, should not be entertained at all, however low the risk.

Fairness.   The extent to which the options affect different people, or groups or segments of society will also have to be examined. Often those affected by a regulation may not share the benefits. For example, adoption of a particular option may affect small firms disproportionately when compared with large enterprises.

Stage 3: Taking Decisions

This is a key stage. It first of all requires a judgement that enough information has been collected and analysed to be able to reach a decision. This will avoid what is known as 'paralysis by analysis' setting in, where the need for additional information is used as an excuse to avoid or postpone the adoption of a decision. All the information gathered in the previous stages needs to be reviewed against a decision-making framework that is explicit and transparent. Success in selecting the most appropriate and workable option for managing the risks will be predisposed by:

- Doing the right things in the first two stages by ensuring that interested parties affected by a particular issue are content with the quality and comprehensiveness of the information gathered, particularly with the results of the risk assessment. They will have to be happy, for example, about how uncertainty has been addressed and the plausibility of the assumptions made
- Reflecting the ethical and value preferences of individuals and society at large when adopting criteria for integrating in the decision-making process the information obtained in the first two stages

Meeting the above conditions can be a tall order, particularly when parties have opposing opinions based on differences on fundamental values or confine themselves to a single issue. Nevertheless, they are worth striving for. If stakeholders are satisfied with the criteria used and how the decisions were reached, they are more likely to support the decisions and help with their implementation.

Criteria Developed by the HSE. Many criteria have been developed for integrating scientific, economic, cultural, ethical considerations, *etc.*, in the decision-making process, with varying degrees of success. However, the framework developed and published[13] by the HSE, known as the Tolerability of Risk (TOR), has gained considerable acceptance from other regulators and industry. Its strength lies in its ability to apply tests very similar to those which we all apply in everyday life for reaching decisions on whether a risk is so great as to be unacceptable; whether the risk is so small that no further precautions are necessary; or, if the risk falls between these two stages, whether the risk should be incurred, taking account of the benefits or the need to avoid some greater risk and with the assurance that efforts will continue to be made to reduce the risk to a level as low as reasonably practicable (ALARP).

The framework is depicted in Figure 1. The horizontal line at the top represents an upper limit, above which a particular risk for practical purposes is regarded as intolerable whatever the benefit. This is the Health and Safety Commission/Executive (HSC/E) view, which may be overridden in the wider national interest by a public enquiry or expressed decision of Government. Any activity or practice giving rise to a risk greater than this threshold would be ruled out by HSC/E unless it can be modified to reduce the degree of risk below this level. For existing risks above the upper limit, remedial action of some kind has to be undertaken irrespective of cost. For example, the importation of certain substances (*e.g.* β-naphthylamine) into the UK is prohibited under the Control of Substances Hazardous to Health Regulations (COSHH) 1992.

The line at the bottom, on the other hand, represents a lower limit or cut-off point where risks are considered broadly acceptable because they are typical of small risks that do not worry people or cause them to alter their behaviour in any

[13] Health and Safety Executive, *The Tolerability of Risk from Nuclear Power Stations*, HMSO, London, 1992.

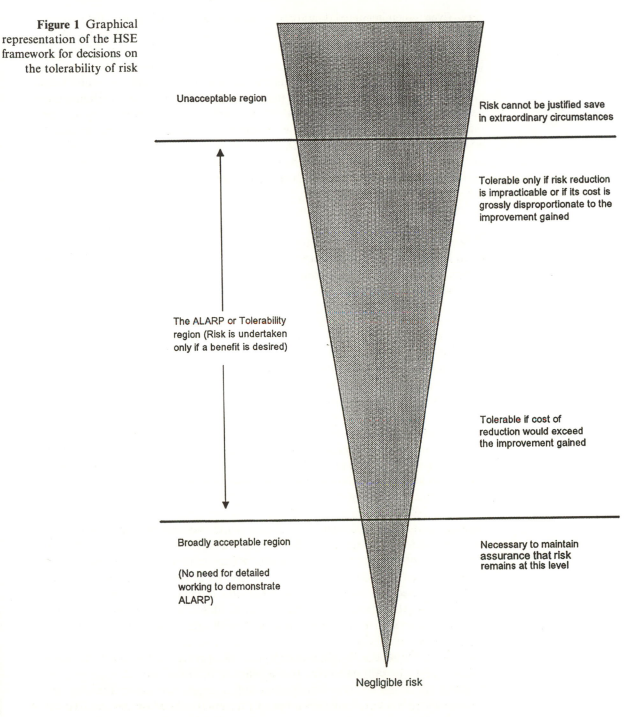

Figure 1 Graphical
representation of the HSE
framework for decisions on
the tolerability of risk

Unacceptable region

Risk cannot be justified save
in extraordinary circumstances

Tolerable only if risk reduction
is impracticable or if its cost is
grossly disproportionate to the
improvement gained

The ALARP or Tolerability
region (Risk is undertaken
only if a benefit is desired)

Tolerable if cost of
reduction would exceed
the improvement gained

Broadly acceptable region

(No need for detailed
working to demonstrate
ALARP)

Necessary to maintain
assurance that risk
remains at this level

Negligible risk

way. Even when incurred, they only result in a very small addition to the background level of risks to which everyone is exposed during their lifetime.

In between the two lines is a region known as the 'tolerability region', where a balance has to be struck and where society will tolerate risks in order to secure social and economic benefits. Benefits for which society generally tolerates risks typically include local employment, personal convenience or the maintenance of general social infrastructure such as the production of electricity or the maintenance of food or water supplies. Society's tolerance of risks is buttressed by an expectation that people will be told the nature and level of risks; that the risks are being controlled as low as is reasonably practicable; and that the risks are periodically reviewed to ensure that controls are both properly applied and also take account of changes over time, for example, the availability of new options for reducing or eliminating risks due to technical progress.

Tolerability Limits.   The dividing lines between the unacceptable/tolerable and tolerable/broadly acceptable regions must reflect society's values at large and will depend on the nature of the hazards and the detriment they could give rise to. The HSE has proposed, as a rule of thumb, that for workers a risk of death of 1 in 1000 per year should be the dividing line between what is tolerable for any substantial category of workers for any large part of a working life and what is unacceptable for any but fairly exceptional groups. For members of the public who have a risk imposed on them 'in the wider interest', the HSE would set this limit an order of magnitude lower at 1 in 10 000 per year.

At the other end of the spectrum, the HSE believes that an individual risk of death of 1 in 1 000 000 per year for the public (including workers) corresponds to a very low level of risk and should be considered as broadly acceptable.

In addition to those levels of individual risks the HSE has suggested[11] that the chance of an accident causing societal concerns should be less than 1 in 1000 per year and possibly less than 1 in 5000 per year, particularly for accidents where there is some choice whether to accept the hazard or not, *e.g.* the siting of a major chemical plant next to a housing estate.

Basis for Limits.   The choice of the above figures is essentially a policy decision based on the HSE's experience. Moreover, the figures are not intended to be strait-jackets to be applied rigidly in all circumstances. The upper and lower limits were determined by analogies with high-risk industries generally regarded as well regulated and taking account of people's voluntary acceptance of risks in particular situations; with the general levels of risk at work that people accept for a personal benefit (such as pay); and with risks that they usually regard as negligible in their walk of life.

On the other hand, the limit for the level of risk giving rise to societal concerns is based on an examination of the prevailing levels of risk that people are prepared to tolerate from the hazard of a major accident affecting the surrounding population, *e.g.* as shown by the survey carried out by the HSE on the potential for a major catastrophe occurring at the industrial installations at Canvey Island on the Thames or the predicted annual chance of an aircraft crash in the UK killing 500 or more people.

Stage 4: How the Decisions are to be Implemented

Adopting quality regulation also requires consideration of how the decisions are to be implemented, since it is futile to take decisions unless they are carried out. Regulators have a range of instruments available to them such as education, information, assistance, guidance, persuasion, promotion, economic incentives, Approved Codes of Practice, regulation and enforcement.

As previously mentioned, implementation very often will take the form of mandatory duties, requiring those who are able to do something about preventing or minimizing the risks to assess and manage them through the introduction of appropriate control measures. In such cases, the HSE[14] and many other regulators have made clear that risk management involves:

- Drawing up well-considered and articulated safety policies where responsibilities are properly defined and allocated and organizational arrangements set out to ensure control and promote cooperation, communication and competence
- Having a plan for taking action, looking ahead and setting priorities for ensuring that risks requiring most attention are tackled first
- Adopting a culture that disposes everyone involved in the implementation to give their best. In the workplace, for example, this means enlisting the co-operation and involvement of all the workforce by getting a commitment, which permeates every level of the organization, to adopt and work to high health and safety standards
- Setting up procedures for monitoring and evaluating progress, *e.g.* by identifying potential indicators for evaluating how far the control measures introduced have been successful in preventing or reducing risks

Stage 5: Evaluating the Effectiveness of Action Taken

Finally, adopting quality regulation would not be complete without *ex post* procedures to establish:

- Whether the actions taken resulted in what was intended
- Whether any modifications are needed to the risk management plan
- How appropriate was the information gathered to assist decisions for action, *e.g.* the methodologies used for the risk assessment and the cost–benefit analysis, assumptions made, ethical and societal concerns, *etc.*
- Whether improved knowledge and data would have helped to reach better decisions
- What lessons could be learned to guide future regulatory decisions, improve the decision-making process and create greater trust between regulators and those affected or having an interest in the risk problem

Evaluation will inevitably be an ongoing process and, like any other evaluation

---

[14] Health and Safety Executive, *Successful Health and Safety Management*, HMSO, London, 1997.

process, needs to be planned. Since there might be some time before the full impact of risk reduction measures can be monitored, the evaluation might first focus on success in implementing the risk management plan before concentrating on the success of the decisions as a whole.

5   Establishing Networks for Achieving More Effective Liaison Between Departments

The responsibility for regulating risks is shared by several Departments and by regulators with a large degree of autonomy. However, although risks usually do not have unique and separate effects on receptors, the boundaries of responsibility between Government Departments often appear to assume that they do. For example, a pesticide has the potential to affect workers, the public and the environment through various pathways such as the air we breathe or the food we eat or drink, yet responsibility for safe use has to accommodate to traditional Departmental boundaries. In such circumstances, there is always a danger that Departments or regulators may adopt too narrow an outlook in regulating the risk within their area of responsibility. This problem has been tackled in two ways.

First, successive Governments have adopted a policy of merging and centralizing the authorities for regulating risk, as exemplified by the creation of the Health and Safety Executive (HSE) in 1974 and the subsequent transfer of responsibility to the HSE for the regulation of Offshore Installations and the Railways. More recent examples are the establishment of the Environment Agency and the announced intention to establish a Food Standards Agency.

There are limits, though, to the extent to which mergers can be carried out. It has been suggested, for example, that there could be merit in creating a single Department or regulator that would have responsibility for regulating all risks whether to man or the environment by whatever mode the exposure to the risks arose. Such a Department or regulator could have responsibility for regulating risks arising from our lifestyle, the air we breathe, the food we eat and drink, our activities at work and the waste we generate. However, although at first sight this option might look attractive, it is hardly a practical solution. The resulting Department or regulator would be so large that it would be virtually impossible to manage and the option would not fit well with our existing parliamentary system. In short, it would be unlikely to lead to better integration of policies.

Additionally, there has been a convergence in the way risks are regulated by Departments. Self-regulating regimes are becoming more and more common with an emphasis placed on those who create hazards to assess and manage the risks that they entail.

The result is that there is now a considerable commonality of approaches between Departments and regulators. Nevertheless, Departments are aware that the Government's approaches to managing and regulating risks have not been developed systematically in Government from the centre but have evolved over time within Departments. There are areas where greater coherence and consistency would lead to tangible benefits. This was the conclusion reached by two reviews which examined the use of risk assessment within Government Departments[15] and their approaches to regulation.[16]

The Inter-Departmental Liaison Group on Risk Assessment

It was such considerations that led to the setting up of an informal committee of those involved in policy development on risk issues known as the Inter-Departmental Liaison Group on Risk Assessment (ILGRA). This committee meets regularly to share views and experience of risk assessment matters and reports regularly to Ministers. In the first of such reports it has identified a number of areas where steps could be taken to achieve greater coherence in policies and practices between Departments. These include:

(i) The adoption of more consistent methodologies to ensure that:

- Different regulators do not adopt conflicting approaches in relation to the same activity
- Departmental frameworks for deciding what risks are unacceptable, tolerable or broadly acceptable can logically dovetail with each other

(ii) Development of methods for ranking risks. One of the greatest changes emerging for Government, as exemplified by several departmental reports,[17,18] is the question of what risks should be given priority, given their perception by the public as well as their objective levels and the availability of technology for preventing or eliminating them.

(iii) Examination of common approaches for conducting cost–benefit analyses. How should the costs and benefits of a regulatory option be balanced, taking into account such issues as differences in the way they are distributed in the future?

(iv) Sharing data and information about risks. For example, most Departments collect, compile and update facts and data on risk which are stored in a number of databases. Yet, for various reasons, some administrative and others technological, they cannot be readily cross-accessed. The decision by the Government to install an Intranet network between Departments should help.

ILGRA Subgroups

ILGRA has established two subgroups to assist it in its work. The first one on 'Methodology for Setting Safety Standards' is chaired by Her Majesty's Treasury

---

[15] Health and Safety Executive, *Use of Risk Assessment Within Government Departments*, Misc 038, HSE Books, Sudbury, Suffolk, 1995.

[16] Her Majesty's Treasury, *The Setting of Safety Standards—A Report by an Interdepartmental Group and External Advisers*, Public Enquiry Unit, HM Treasury, London, 1996.

[17] Health and Safety Executive, *Occupat. Health Rev.*, 1994, **48**, 24.

[18] Department of Health, *Health of the Nation—A Strategy for Health in England*, Cm 1986, HMSO, London, 1992.

and has experts from outside Government as members. This sub-group is looking at the economic foundations of Departmental approaches to the setting of safety standards in the light of the Treasury report[16] already mentioned above. It is proceeding by examining case studies provided by Departments to illustrate their approach to the regulation of risk.

The second sub-group is chaired by the Ministry of Agriculture, Fisheries and Food and has been set up in conjunction with the Research Councils to develop a joint strategy on toxicological risk assessment. In particular, the sub-group will review current UK and overseas practice and develop a research strategy aimed at generating new approaches to the estimation of risk to health from toxic substances in the light of new scientific techniques, such as *in vitro* experimentation using human tissues, molecular modelling, computer simulation and the use of human biomarkers.

Research

ILGRA also provides a means for enabling Departments to collaborate on research in a number of important areas. These include the ranking of risk, the valuation of safety benefits and risk communication.

The work of ILGRA dovetails well with the recommendation of the Technology Foresight Steering Group in their 1995 report[19] that Departments give adequate attention to ensuring that risk and risk management should be a generic science and technology priority. As is apparent from the above, a lot of its activities and those of its sub-groups are concerned with examining the development of methods and protocols in relation to risks arising from new technologies and fostering long-term thinking on risk issues.

6 Improving Systems for Communicating Risk Information to the Public

Departments have always accepted[15] that risk communication should be an integral part of the way in which Government manages risk in its various roles as an investor (*e.g.* for determining how much public money it is desirable to invest in education, the health service, transport); a regulator (*e.g.* for determining whether it should regulate or not); an enforcer (*e.g.* for selecting targets and determining the frequency of checks for determining that the law is being complied with); and an educator (*e.g.* in providing guidance for complying with the law or on good business practice or healthy lifestyle choices). For example, it is central to the HSE's fulfilling its own mission statement to ensure that risks to people's health and safety from work activities are properly controlled.

However, under the aegis of ILGRA, Departments are examining their current procedures. The reasons for this are:

- A growing awareness that what has worked well in the past may no longer be appropriate. For example, in the past, Departments have often relied on

[19] Cabinet Office, Office of Public Service and Science, *Progress through Partnership—Series of 15 Reports from the Technology Foresight Programme Sector Panels*, HMSO, London, 1995.

expert opinion as a source of reassurance, particularly for those risks where people cannot judge for themselves what are the risks (a feature associated with many of the newer hazards arising from industrialization, *e.g.* the risk posed by traces of pesticides in food). However, the trust placed in expert opinion as a source of reassurance is being continually eroded, particularly for those issues where the mass media have exposed controversies surrounding such opinions

● A recognition that current procedures may be inadequate to take account of increased media attention. The rise of information technology can lead to the public response to risk being amplified or attenuated, depending on how the risk interacts with psychological, social, cultural and institutional processes

● Departments and regulators are increasingly being placed in a position where they have to demonstrate their authority, credibility and efficiency and are aware that good and effective means of communicating with their stakeholders are effective weapons in responding to such pressures

Departments agree that, to succeed, risk communication must be essentially a two-way or interactive process which aims to:

● Promote understanding and inform the public about health and safety both in normal and in emergency situations

● Enable the informed views of the public to be taken into account in Government decision-making

● Improve mutual understanding of public and Government attitudes to policy making about health, safety and the environment, *e.g.* by identifying the reasons for differing opinions and seeking ways to arrive at decisions which can be accepted with minimum dissent by the people involved

● Develop effective alternatives to direct regulatory control

● Influence human behaviour in a positive manner

These principles and the findings of the research on risk communication mentioned above will be integrated in guidance that the ILGRA has commissioned on behalf of all Departments on this subject.

7 Strengthening the Links with Industry, Non-Governmental Organizations and Academic Institutions

The regulatory apparatus within Government has always had a limited capacity for evaluating all health, safety and environmental risks and identifying workable options for tackling these risks. Departments have often relied upon contributions from industry, academia, pressure groups and non-governmental organizations to assist them in those tasks. Such contributions have often proved to be mutually beneficial; for example, by providing the Government with access to the best scientific expertise available; for accessing the views of stakeholders on specific issues without carrying out expensive surveys; and for influencing research at the formative stage so that investigations are targeted to achieve the maximum benefit. Such co-operation is bound to grow as policy questions hinge increasingly on complex scientific and/or controversial questions.

However, Departments and regulators are aware that such co-operation is not without its dangers. They must tread carefully to ensure that this does not lead to a lack of trust as would inevitably occur if they were perceived as having been 'captured' in the course of the co-operation.

8 Concluding Remarks

Government has reached a cross-roads in the regulation of risk. This has traditionally been tackled on a Departmental basis according to approaches rooted in the culture of the Department concerned and the wealth of experience amassed from regulating over the years. Though there are great similarities in these Departmental approaches, there are also notable differences.

Several factors—some scientific and other social, political and international—are causing Departments to converge in their approaches to the way they regulate and communicate on risks. Departments are also coming under more pressure to adopt quality regulation and to be open in their processes for taking decisions on risks. In this context the framework adopted by the HSE for deciding what risks are unacceptable, tolerable and negligible has gained a large measure of acceptance.

The Inter-Departmental Liaison Group on Risk Assessment is playing a major role in ensuring that Departmental approaches are consistent and in commissioning research which would help to achieve this objective. As such, it fits very well with the quest by Government for improved regulatory policies that will help ensure both wealth creation and quality of life for this and future generations.

Pollution Risk Management

SIMON HALFACREE*

1   Introduction

The Environment Agency was formed in April 1996 by the amalgamation of the National Rivers Authority, Her Majesty's Inspectorate of Pollution, the Waste Regulation Departments of Local Authorities and the Waste Technical Division of the Department of the Environment. The Agency is responsible for England and Wales; separate arrangements are in place for Northern Ireland and Scotland. The Agency acts under a variety of statutes[1-3] with responsibilities in the area of environmental protection, water resources, flood defence, fisheries, conservation, navigation and recreation. Of particular relevance to the subject of this article are the Agency's duties to undertake pollution prevention and control in relation to all components of the environment: air, land and water. The Environmental Protection Department comprises of Waste Management and Regulation, Process Industries Regulation, Radioactive Substances Regulation and Water Quality functions.

The Agency is required to undertake its duties with consideration of the need to achieve sustainable development 'taking into account the likely costs and benefits' and this is reflected in the Agency's vision for 'A better environment in England and Wales for present and future generations'. One of the main objectives of the Agency is to 'protect and improve the environment as a whole by effective regulation, by our own actions and by working with and influencing others . . .'. As one way of achieving these aspirations, the Agency is at the forefront of developments, *via* its extensive applied R&D programme, so that it has the best possible techniques at its disposal. An important set of techniques, developed in recent years, are those employing the principles of risk assessment and risk management. The Agency has developed risk-based methods in many of its areas of responsibility, but the subject of this article is the assessment and management of pollution risks to the environment.

There are three main routes by which potential pollutants may enter the environment: *via* routine authorized discharges, by accidental releases and by

*The views expressed in this article are those of the author and, except where the context indicates, are not necessarily those of the Environment Agency.

[1] *The Environmental Protection Act 1990*, HMSO, London, 1990.
[2] *The Water Resources Act 1991*, HMSO, London, 1991.
[3] *Environment Act 1995*, HMSO, London, 1995.

diffuse inputs, for example from the use of pesticides. Risk management techniques are applicable to the management of risk of pollution from all of these routes, although they are more relevant to accidental releases and, consequently, accidental releases are given greater emphasis in this article. The key elements of a risk-based approach are consideration of the following five questions, as they apply to the activity being assessed:

- What can go wrong?
- How often?
- What are the consequences?
- Is the combination of frequency and consequence acceptable?
- How can the combination of frequency and consequence be reduced to an acceptable level?

In answering these questions, it is important to get the right balance between the complexity of the assessment process and the level of protection which can be afforded to the environment by any risk reduction measures which might be identified. In other words, costs and benefits must be proportionate. In order to achieve this aim, a number of levels of risk assessment of increasing complexity are employed, with progression to the more complex levels depending on demonstration of significant risk by the simpler techniques. By adopting this approach, the lower risk activities are 'screened out' and do not require more complex assessment. In line with the precautionary principle, where information is incomplete, progression to a more complex assessment is generally required. Where there is insufficient information to make a judgement, then it may be necessary to implement a set of standard pollution prevention measures which will, of necessity, be precautionary. At each level of assessment, costs and benefits are taken into consideration. The protocol for the cost–benefit element of the decision making process is presented diagrammatically in Figure 1.

Although, in some circumstances, the Agency may carry out the simpler 'screening' techniques on behalf of industry, it will usually be up to industry to provide an appropriate assessment in support of their application, for whichever formal permission they require. The Agency does not prescribe what techniques should be used, but seeks to provide 'model techniques' and general guidance that, if followed, will ensure an appropriate level of assessment by industry. However, the assessment will be subject to technical audit by Agency staff and the results will be compared to relevant criteria. Agency staff need to be involved at the earliest stages of an assessment and not simply presented with the results of a completed study, to ensure that the proposals are in line with what is required for a particular site.

The main benefit of risk assessment is its ability to target the attention and resources, of both the regulators and industry, to the potentially most damaging activities. In this context, resources means both the expertise and the finance which might be required to undertake the assessment and introduce any risk reduction measures which might be needed. In addition, the adoption of risk-based techniques promotes a systematic approach, so that the likelihood of overlooking a potential source, or mechanism, of pollution is much reduced. A further benefit is the consistency which such an approach generally requires, so that different

Figure 1 Risk assessment
cost–benefit protocol

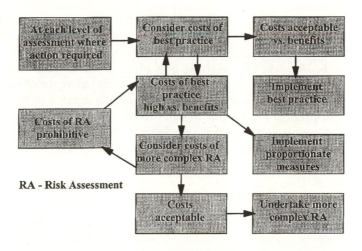

RA - Risk Assessment

industries, in different parts of the country, have a similar approach and standard of assessment that they must achieve. Where the approach does result in a greater resource requirement, it is because of the risk that the site or activity poses to the environment.

In this article, emphasis is placed on the methodology which has been developed to support an application for a 'Protection Zone' on the River Dee, as this is an area of the Agency's responsibility where a complete set of techniques and criteria has been developed and tested, by being subject to a Public Inquiry. The development of a methodology for protection of potable water supplies, using risk management techniques, has been the first priority and the techniques are subsequently being extended to protection of the environment as a whole. Other areas are described in less detail, but sufficient to provide an indication of the range of risk-based techniques which are being developed. There are common elements to many of the techniques which are described but they all include, to some extent, the following stages:

- Hazard identification
- Estimation of release frequencies
- Estimation of consequences, sometimes by modelling
- Comparison of risk against defined criteria

There is overlap between some of the methods described and this is partly historical and partly to ensure that appropriate techniques are available to suit the size of problem and the level of expertise of the assessor. In the future these techniques should be rationalized to ensure that a comprehensive and coherent set of techniques is available, to protect all elements of the environment.

2 Potable Water Risk Management

The Water Industries Act 1991 (Section 68)[4] requires the water undertakers 'to supply only water which is wholesome at the time of supply . . .', which is usually

[4] *Water Industries Act 1991*, HMSO, London, 1991.

interpreted as being water free from contaminants that could affect its palatability or the health of the consumer. The Environment Agency is obliged to have regard to the statutory duties of the water companies. 'It shall be the duty of the Agency, in exercising any of its powers under any enactment, to have particular regard to the duties imposed by the virtue of the provisions of . . . the Water Industry Act 1991, on any water undertaker or sewerage undertaker that appears to the Agency to be or to be likely to be affected by the exercise of the power in question' (WRA 91, Section 15). The protection of sources of water from accidental contamination is ideally suited to a risk-based approach.

Water Protection Zones

Section 93 of the Water Resources Act 1991 allows the Environment Agency to apply to the appropriate Secretaries of State for the designation of Water Protection Zones (WPZ), where activities can be controlled or prohibited. The powers under this section, to require pollution prevention measures, go beyond those available under other pollution control legislation. The Welsh River Dee has for some time been identified as the highest priority for protection zone status in England and Wales and in 1994 an application for designation of the freshwater part of the catchment, as a protection zone, was made to the Welsh Office and the DoE, leading to a Public Inquiry in March 1995. The Dee provides about 2 million people in NE Wales and Merseyside with their drinking water supplies and the potable water abstraction points are located downstream of numerous industrial premises. There are frequent pollution incidents on the Dee and a serious incident in 1984, where the river was polluted with phenol, led to wide distribution of contaminated water and concern about possible health effects. The financial implication of the 1984 incident were considerable, with estimated total costs of approximately £20 million. Alternative solutions to the problem of pollution of the potable intakes, such as bankside storage, were either impracticable or more expensive than establishing a protection zone. The protection zone approach is compatible with the 'polluter pays' principle.

The Dee has been seen as a pilot case for implementation of this legislation and there may be applications for protection zones on other rivers, depending on the lessons learned on the Dee. However, it is unlikely that protection zone powers will be widely implemented.

Prior to the protection zone application, the National Rivers Authority (now part of the Environment Agency) embarked on a programme of R&D, to develop a methodology to support the application at the Public Inquiry and form the technical basis for the implementation of the protection zone order. The purpose of the programme of R&D was to develop a set of techniques that addressed all parts of the process, by which the risks of industrial operations could be determined to be acceptable, or not. This included:

- Obtaining data on the identity and quantity of chemicals stored in the catchment
- Estimating the size and likelihood of chemical spillages from individual sites

- Predicting the consequence of such spillages
- Determining whether the combination of frequency and magnitude of the consequences is acceptable

The details of the methodology are described in more detail in the following section.

Methodology for Implementation of Protection Zones

Catchment Inventories. In order to make an assessment of the risks in a catchment and understand how they should be managed, it is important to have a list of the hazards. In this context, the hazards take the form of stored chemicals, or chemicals in reaction vessels. A catchment inventory is essential in order to undertake initial screening, prior to more complex assessments.

When compiling an inventory it is necessary to apply a quantity cut-off, to avoid including prohibitively large amounts of data on very small quantities of substances, which will usually present a negligible risk. In order to determine the appropriate quantity at which to apply a cut-off, consideration must be given to the most toxic substances which might be found in a catchment and the initial dilution which might be afforded if they were to be accidentally released. The decision on the cut-off quantity is a compromise, where a balance between collecting a manageable amount of data and exclusion of potentially significant chemicals must be reached.

PRAIRIE (*Pollution Risk from Accidental Influx to Rivers and Estuaries*)

PRAIRIE[5] is a piece of PC-based software, running under Windows, that facilitates the assessment of the consequence element of the risk management process in rivers and accepts the input to data on likely spill frequencies from a spill risk assessment. The main elements of PRAIRIE are:

- DYNUT—a one-dimensional aquatic dispersion model
- A substance database, including physico-chemical data and toxicity information
- A hydrological database, containing information on river flows and weirs
- A tabular and graphical output facility
- A standards database, containing limits relevant to potable water and the environment

DYNUT is a one-dimensional model, which is applicable to non-tidal rivers or streams. It uses a finite difference solution scheme to solve the advection–dispersion equation in each of a number of partitions into which the river is divided. The main data requirements are information on distances, flows, velocities and the

[5] L. S. Fryer and T. Davies, *A Guide to Risk Assessment Methodologies*, R&D Note 371, National Rivers Authority, Bristol, 1995.

Figure 2 DYNUT model concept

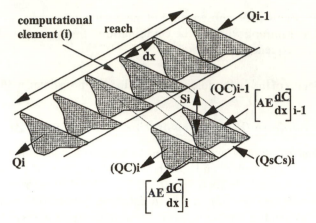

longitudinal dispersion coefficient. The model is capable of representing the situations where a chemical is spilt into a tributary, which subsequently enters the main river.

As an option, DYNUT is able to predict the reductions in river concentration caused by various processes which may occur. It has the capability to simulate the following processes:

- Volatilization
- Photolysis
- Oxidation
- Adsorption to sediment
- Hydrolysis

The mechanisms by which the concentration of pollutant may be reduced are known as 'sinks' and the model can be employed with sinks switched 'on' or 'off', in one of two modes. If the sinks are switched on, then there are various additional data requirements, although the software contains defaults, if information is not available. The DYNUT model concept is illustrated in Figure 2, where Q = flow, C = concentration, E = diffusion coefficient, A = cross-sectional area and S = sources and sinks. The figure represents the division of the river reach being modelled into a number of computational elements. A single computational element is shown separately to indicate the calculations which are undertaken for each element, to determine the fate of the pollutant.

At the present time, PRAIRIE contains data on 250 substances which have been identified as being of high priority, as a result of studies on the Dee and other catchments. Hydrological data are currently incorporated for a limited number of catchments, but additional catchments can be added.

PRAIRIE can be employed in a number of modes, depending on what is appropriate to the particular level of study being undertaken. The simplest mode is the deterministic mode, in which PRAIRIE is used to model a single chemical under defined conditions of source-term and river flow. There are three graphical options for displaying the results:

Figure 3 Simple PRAIRIE
screening output
(HUMSNARL = Human
Suggested No Adverse
Response Level)

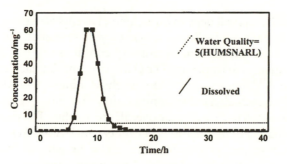

- Three dimensional (time, distance, concentration)
- Concentration *versus* time at a chosen downstream distance
- Concentration *versus* distance at a selected time

The concentration *versus* time, or distance, plots can be overlaid with selected water quality standards, to allow rapid assessment of the possibility and likely extent of their being exceeded for a particular spill scenario. A typical application would require consideration of the concentration *versus* time output at a potable water intake, selecting the water quality standard appropriate to protect the use of abstracted water for potable supply. Such an assessment would form part of a screening exercise, with realistic worst case assumptions being made about the 'source term' and the river flow. If the rapid loss of the total inventory of a chemical, from a site under consideration, could not possibly result in the relevant standard being exceeded, then the chemical under question is excluded from further assessment. This demonstrates how PRAIRIE can be used to 'screen out' low risk sites and so target resources to those sites where more toxic chemicals are stored in a particular quantity and location, which combine to generate the potential for a significant impact in the event of an accidental spill. Figure 3 shows a typical output graph from a deterministic run, showing the predicted pollutant concentration compared with the relevant water quality standard for protection of potable water supplies. The distance chosen for the plot would be that representing the location of the nearest potable water abstraction point.

In its probabilistic mode, PRAIRIE can model a number of different chemicals, each with up to five different spill scenarios, and take account of the probability of particular river flows at the time of release, by using data contained in a hydrological database. The output takes the form of a plot of frequency *versus* multiples of a selected water quality standard, which is compatible with the criteria which have been developed for assessing the results (see below).

Spill Risk Assessment. The part of the risk management process which is most specialized is the assessment of the probability of various release scenarios occurring. For this reason the adopted risk management strategy for the WPZ avoids requiring spill risk assessments wherever possible. This is achieved by 'screening-out' chemical stores which, even in a realistic worst-case spill scenario, cannot result in concentrations of concern being exceeded in the receiving water. When a spill risk assessment is required, it is necessary to identify all of the possible scenarios which might result in the substance entering the watercourse.

In many cases this will depend on a chain of events occurring, with a probability attached to each stage. Some scenarios will represent the situation where a substance has escaped from its primary containment, but does not ultimately enter the aquatic environment. The chain of events for each scenario will include the failure of containment devices, such as bunds. In order to calculate the likelihood of each alternative combination of events occurring, generic failure rate data are used to calculate the overall probability. The result of this analysis is a series of possible combinations of events which result in spills of varying sizes, each with a related probability. These can be entered into PRAIRIE in its probabilistic mode, to produce the frequency *versus* concentration curve for assessment against the relevant risk criteria.

Potable Water Risk Criteria. The purpose of the risk criteria which have been developed[6] is to provide a benchmark against which to measure a particular site, as a means of deciding whether the risks to the potable water supply from hazardous substances are acceptable. The results from a quantified risk assessment can be compared with the criteria to determine whether action is necessary. As risk comprises a component describing the probability or frequency and one describing the magnitude of the consequences of an accident, the risk criteria need to incorporate both of these factors. Potable water abstraction points along a river may be at risk from several upstream installations, so that the overall risk of unacceptable water entering the abstraction point is the sum of the risks posed by the individual facilities. This means that the same overall risk will be experienced at the abstraction point, both from a relatively large number of facilities, each posing a low risk, and from a few facilities, each of which poses a relatively large risk. However, in order to be able to carry out risk assessments on individual sites it is necessary to consider each facility separately and, therefore, the criteria to help evaluate the acceptability of risks at the abstraction point were derived in the form of risks from individual industrial operations upstream of potable water intakes.

The criteria have been based on the philosophy that a plot of frequency *versus* severity contains three regions:

- A region associated with low risks which are considered 'acceptable'
- A region of high risks which are considered 'intolerable'
- An intermediate region, where risks may be 'tolerable' providing that they are as low as reasonably practicable (ALARP)

The risk criteria also needed to meet the following requirements:

- Applicability to river systems containing potable water abstraction points
- Compatibility with the output of a risk assessment tool such as PRAIRIE
- Capability of allowing for risks posed by a variety of accidental releases
- Applicability to accidental releases from an individual site

In considering the possible effects that the accidental release of a chemical

---

[6] L.S. Fryer and W. Nixon, *Development of Risk Criteria for the Protection of Potable Water Abstraction Points*, R&D Note 369, National Rivers Authority, Bristol, 1995.

might have, it is necessary to consider a range of effects, from minor, such as taste and odour complaints, through to health effects, which in some circumstances might be potentially severe. In the absence of data which could be related to specific effects for each chemical, it was necessary to use some other pertinent data which are available. The UK Water Research Centre (WRc) has developed Suggested No Adverse Response Levels (SNARLS), which are concentrations in potable water which can be consumed safely for a specified period, usually 24 hours or seven days.[7] SNARLS also take account of organoleptic properties, which could result in consumer complaints but which are not necessarily toxic.

In order to consider the more severe effects, in the absence of data on humans, mammalian toxicity data were used. The data employed were the LC_5 and the LC_{50}, being the concentrations which would be lethal to 5% and 50% of an exposed population, of a particular species, in a defined period of time. The LC_5 represents the concentration where a 'dangerous dose' would be received by members of the population consuming the water. The LC_{50} represents the concentration where consequences would be extremely serious and, although there is a tendency not to contemplate such scenarios, they could theoretically occur and must be taken into account in any risk criterion scheme.

In order to derive frequencies to accompany the range of concentrations of interest, reference was made to existing situations deemed to be unacceptable and to precedents which had been set in other risk management fields. For the lower concentrations, reference was made to the current position on the River Dee, where the number of existing incidents is tolerable but not acceptable. Following consultation with the Dee Steering Committee, which includes representatives of all of the water companies, it was decided that the frequency of exceeding the 7 day SNARLS and 24 hour SNARLs should both be reduced by a factor of 10 from existing levels to define the acceptable frequency. The tolerable frequency was set at the current frequency. With regard to the higher concentrations, reference was made to precedents which had been put forward for broadly equivalent consequences in the field of human safety and acceptable and tolerable frequencies were derived for both LC_5 and LC_{50}.

The criteria which were put forward to support the Dee Protection Zone application therefore take the form of a graph of frequency against concentration having three separate zones: acceptable, As Low As Reasonably Practicable (ALARP) and intolerable (Figure 4). The main purpose of the criteria is to guide discussions with industry and not to act as part of a rigid decision-making apparatus.

Some other rivers, which have similar problems to the Dee, are having comparable techniques applied in a less formal way whilst experience is gained of the practicalities of a protection zone approach.

Integrated Pollution Control Sites in Protection Zones

A different approach was taken with the larger industrial sites in the proposed protection zone which come under the Integrated Pollution Control provisions

[7] G. O'Neill, *Acceptable Frequency for the Consequences of Accidental Release of Toxic Chemicals into a River used for Potable Water Abstraction—Toxicity Considerations: Phase 2*, Project Record 292/2/W, National Rivers Authority, Bristol, 1994.

Figure 4 Potable water risk criteria

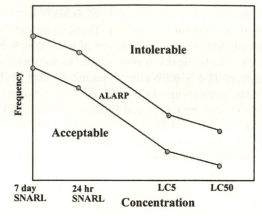

of the Environmental Protection Act, because the Agency has more powers on these sites and they were therefore exempted from protection zone control. However, a methodology was developed specifically for IPC sites which has many similarities. As well as ensuring an equivalent level of protection for potable water supplies from the risks posed by these sites, the methodology which was developed also considered protection of the general aquatic environment and risks of pollution of air and land. The methodology for IPC sites in protection zones has three steps:

- Hazard identification
- Screening assessment
- Likelihood consequence assessment

The technique developed will need to be further evaluated, but may in due course form the basis of a risk assessment which might be required, in certain instances, as part of the IPC Authorization process, although it is likely that this will be overtaken by the IPPC Directive requirements. The Integrated Pollution Prevention and Control (IPPC) Directive[8] will have major implications for a risk management approach and must be implemented in the UK by October 1999. It is estimated that 9000 installations in the UK will come under IPPC and the Directive has specific requirements relating to the prevention of accidents and will require permitting of sites, taking into consideration the measures which are in place to prevent accidents. At the time of writing, the details of risk assessment requirements under IPPC are still being developed.

3   General Environmental Risk Management

Substance-based Risk Management

Risk assessment can be approached on a site or activity basis, or alternatively from a substance viewpoint. Regulations on the Risk Assessment of New and

---

[8] Council Directive 96/61/EC, *Integrated Pollution Prevention and Control*, Official Journal of the European Communities, Brussels, 1996.

Existing Substances are in place and provide a mechanism for looking at the nature and quantities of substances used in the UK. Where there are significant quantities of hazardous substances in use, then appropriate precautionary measures can be identified or alternative chemicals considered. The risk assessment of new substances provides an opportunity to decide whether the introduction of a new substance can be justified from the point of view of the risk which it poses to the environment. Dangerous substance controls are based on the premise that for some substances the consequences of release are such that no release is acceptable. Further consideration of substance-based risk management is outside the scope of this article.

Risk Management of Routine Releases

Routine releases to the environment made under a consent or authorization are controlled so that they should not cause pollution. The means by which this objective is achieved differs between the major items of pollution control legislation, the Environmental Protection Act 1990 (EPA90) and the Water Resources Act 1991 (WRA91).

EPA90 was the route for introduction of Integrated Pollution Control (IPC), which applies to the potentially most polluting industries. IPC is designed to provide an integrated approach, by requiring industries to obtain an authorization to operate a prescribed process. The authorization deals with discharges to air, water and land, and in deriving the authorization conditions the principles of Best Practicable Environmental Option (BPEO) and Best Available Techniques Not Entailing Excessive Cost (BATNEEC) are applied. The legislation makes a distinction between the larger 'Part A' processes and smaller 'Part B' processes which are controlled by Local Authorities. BATNEEC is used for:

- Preventing the release of substances, prescribed for any environmental medium, into that medium, or, where this is not practicable, reducing the release to a minimum and rendering any release harmless
- Rendering harmless any other substances that might cause harm if released into any environmental medium

The determination of consents to discharge to water, under the Water Resources Act 1991, is usually based on a realistic worst-case approach. It is not explicitly risk-based, but takes account of risks by considering the worst case in relation to effluent flow, composition and available dilution. Consents are usually designed to protect the environment from the maximum permitted load discharged into a river under low flow conditions. Environmental Quality Standards (EQSs), both statutory and non-statutory, are used as the criteria for acceptability.

Pollution Prevention Guidance

An alternative to a risk-based approach is one which requires precautionary measures to be implemented, regardless of the risk posed. This has the advantage of dispensing with the need to undertake the risk assessment, but can be less cost-effective as it is not targeted. In order to provide advice to industry as to what precautionary approaches are appropriate in particular circumstances, the Agency has produced a variety of standard guidance, including a series of Pollution Prevention Guidance Notes, outlining requirements in particular situations. This guidance is often relevant to smaller businesses, where it is more cost-effective to install precautionary measures than to use resources in undertaking risk assessments.

'Rapid Risk Assessment'

A series of techniques for protection of the aquatic environment were developed within the Agency as practical risk assessment tools that could be used by non-specialists.[9] The idea was to develop an approach which was quick and only required data which were readily available. A proposed set of techniques was trialed across the Agency and a clear need to modify the proposals to include a simpler approach, with minimum data requirements, was identified. The basic level of assessment which was developed recognized the difficulties which operational staff encountered when trying to obtain data for large numbers of sites and takes the form of a 'checklist'. The introduction of such an approach was designed to allow such staff to make decisions in a consistent and systematic way and provide a point-of-entry to the risk management process. The final version of the suite of rapid risk assessment techniques has three levels, which complement the use the PRAIRIE software and do not require use of a computer or knowledge of computer models.

The first stage is known as the 'Initial Hazard Assessment' and requires the hazard to be subjectively assessed against 15 possible issues, divided in to categories of:

- Type of site/activity
- Substances
- Incident history
- Vulnerability of watercourse/groundwater
- Use of watercourse/groundwater

The second level assessment uses a scoring system to reduce subjectivity, but considers the same issues in more detail and with greater emphasis on site management. This level of assessment is known as a 'Primary Risk Assessment' and, like the Initial Hazard Assessment, the need for detailed toxicity and flow information is kept to a minimum.

The third level in the Rapid Risk Assessment suite has greater data requirements than either of the other levels and overlaps with a simple PRAIRIE assessment, but can be undertaken without the use of a computer. This method requires both river flow and toxicity information and its use is therefore appropriate where the primary risk assessment indicates cause for concern.

[9] National Rivers Authority, *Pollution Prevention Manual*, NRA, Bristol, 1995.

Figure 5 Risk assessment protocol

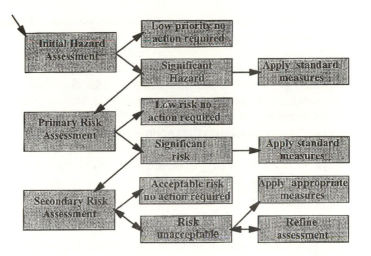

The idea of the hierarchy of levels is that each one acts as a screen to eliminate the sites or activities which cannot cause a problem. Only the sites which appear unsatisfactory in the simpler assessments are considered by the more complex techniques. The hierarchy of assessment levels is represented diagrammatically in Figure 5 and this must be considered in conjunction with the cost–benefit protocol (Figure 1) to ensure that costs and benefits are taken into consideration at each stage of the assessment.

Environmental Risk Criteria

In order to apply risk assessment tools such as PRAIRIE to protection of the general environment, it is necessary to develop criteria applicable to protection of the ecosystem rather than consumers of potable water. An important item of work which has been undertaken to address this requirement, called 'Management of Harm to the Environment', is discussed in Section 5.

Groundwater

The protection of groundwater from pollution needs a different approach from surface waters in recognition of the long-term nature of any pollution and the difficulty of predicting the fate of contaminants. The use of risk assessment in relation to groundwaters falls into one of the three categories:

- Determining vulnerable areas and situations which are at risk (*e.g.* Groundwater Protection Policy[10] and its associated vulnerability maps and source protection zones)
- Assessing new developments to determine the acceptability, constraints or engineering needs to mitigate potential impact (*e.g.* LandSim)
- Regulating existing operations or land/water quality to evaluate the

---

[10] National Rivers Authority, *Policy and Practice for the Protection of Groundwater*, NRA, Bristol, 1991.

activities/contaminants that pose the greatest threat, and prioritizing regulatory actions.

LandSim (Landfill Risk Assessment by Monte Carlo Simulation). The software package 'LandSim' has been developed for the Agency as a tool to help in the assessment of new landfill sites. It forces regulators to think about their decisions in terms of risks to groundwater and encourages discussion and agreement with landfill operators over the key data needs. It promotes consistency without being prescriptive. Similar tools to LandSim are being developed as assessment aids in the related areas of groundwater pollution prevention and remediation and contaminated land management.

BPEO (*Best Practicable Environmental Option*) *Methodology*

Where an IPC process involves the routine release of substances to more than one environmental medium, the Agency needs to determine the Best Practicable Environmental Option for the pollutants concerned.[11] This is a form of risk management, taking account of the possible consequences in each of the media which might be affected, so that an option can be selected which has the least potential for environmental damage. In the 12th report of the Royal Commission on Environmental Pollution, BPEO was defined as 'the option which in the context of releases from a prescribed process provides the most benefit or the least damage to the environment as a whole, at an acceptable cost, in the long term as well as the short term'.

A key element in the assessment of the BPEO is the evaluation of the impact of releases on the environment as a whole. The integrated assessment of the impact of releases on the environment is complex. The effect of a release on the environment will be dependent on many factors, including:

- The amount of each substance released
- The rate of release of each substance
- Other release characteristics
- The physical properties of the released substances
- The chemical properties of the released substances
- The nature of the receiving medium
- Ambient concentrations of released substances already in the environment
- The location of the receptors in the environment, sensitive to the released substances
- The degree of sensitivity of these receptors to enhanced concentrations of released substances

Because of the complexity of taking all these factors into account, a simplified approach has been developed which is based on the direct environmental effects of releases to a particular environmental medium, assessed as the ratio of the process contribution to the Environmental Assessment Level (EAL) for that

[11] Environment Agency, *Best Practicable Environmental Option Assessments for Integrated Pollution Control—Guidance for Operators and Inspectors of IPC Processes*, HMSO, London, 1997.

substance in that medium. The EAL is the concentration of a substance which, in a particular medium, acts as a comparator value which enables an evaluation to be made of the relative environmental effects of different substances. The BPEO assessment also takes Global Warming Potential and the potential for ozone generation into account.

OPRA (*Operator and Pollution Risk Appraisal*)

OPRA is a risk assessment system (see also Section 5 of the Gerrard and Petts article in this volume) which has been developed to assess the risks posed to the environment by IPC sites.[12] OPRA has only recently been introduced and will initially be used to prioritize site inspections by the Agency, although it may be used for a range of purposes in the future. There are a number of factors which influence the risk of a process. The risk level is determined by the balance between the pollution hazard and the operator performance and, for this reason, OPRA consists of two separate parts: the Operator Performance Appraisal (OPA) and the Pollution Hazard Appraisal (PHA). Each part involves the allocation of a rating from 1 to 5 in relation to seven attributes. OPA evaluates operator performance against the following key attributes:

- Recording and use of information
- Knowledge and implementation of authorization requirements
- Plant maintenance
- Management and training
- Plant operation
- Incidents, complaints and non-compliance events
- Auditable environmental management systems

The purpose of the OPA is to evaluate operator performance, which requires consideration of the systems and procedures in place and also whether they are effective in achieving the operator's stated objectives in relation to environmental performance. For each of the OPA attributes, consideration should be given to the following:

- Do the appropriate systems exist?
- Are the systems used as intended?
- Are the systems effective in achieving stated objectives?
- Is there appropriate monitoring and feedback on the system performance?

Each attribute is evaluated on the scale 1 to 5, where 1 represents low performance and 5 represents high performance.

A Pollution Hazard Appraisal is performed by evaluating the following seven attributes:

- Presence of hazardous substances

---

[12] Environment Agency, *Operator and Pollution Risk Appraisal (OPRA) Version 2*, Environment Agency, Bristol, 1997.

- Scale of use of hazardous substances
- Frequency and nature of hazardous operations
- Techniques for prevention and minimization of hazards
- Techniques for hazard abatement
- Location of process
- Offensive substances

PHA is intended to evaluate the overall pollution risk inherent in a process.

OPRA is designed to complement authorizations and support inspections by providing information about IPC processes and their performance against the conditions of the authorization. OPRA is a simple and rapid screening tool which provides a simple characterization of the overall environmental risk from a process. It is not intended for detailed assessment of process risk or operator performance.

The expected benefits of OPRA are as follows:

- Agency resources will be objectively allocated and efficiently employed
- Targeting of inspection effort at high risk/low performance processes may reduce the total number and severity of incidents
- Greater transparency, improved communication, standard approach to evaluation of risk, together with the process 'track record' which the OPRA score will provide, should add to the benefit derived by the operator and the Agency inspector from an inspection visit
- Inspection procedure may become more efficient, by using OPRA as a systematic review of the important factors affecting the risk of the process
- Well operated or low risk processes may receive fewer inspections, reducing the cost to operators of preparing for and hosting inspections

Potential concerns in specific areas would be indicated by a low score in a particular OPA attribute or a high score in a PHA attribute. The scoring system is therefore structured, both to identify the key areas of concern and to illustrate the fundamental concept that the risk level is determined by the balance between the overall operator performance and the inherent risk in the process.

4 Major Accident Risk Management

Following the Seveso incident in 1976, when an industrial accident led to a significant release of a dioxin, the EU introduced a directive to control Major Accident Hazards, requiring consideration of possible effects on both man and the environment.[13] The first Seveso Directive was implemented in the UK by the Control of Industrial Major Accident Hazards (CIMAH) Regulations 1984 and these are due to be replaced by Regulations under the Control of Major Accident Hazards (COMAH) Directive[14] by February 1999.

---

[13] Health and Safety Executive, *A Guide to the Control of Industrial Major Accident Hazard Regulations 1984*, HMSO, London, 1990.

[14] Council Directive 96/82/EC, *Control of Major Accident Hazards Involving Dangerous Substances, Official Journal of the European Communities*, 1996.

CIMAH (*Control of Industrial Major Accident Hazards*) *Regulations*

The CIMAH Regulations are the legal instrument implementing the Seveso I Directive in the UK and are applied by the Health and Safety Executive (HSE), because they are primarily intended to protect safety of humans. In 1991 the Department of the Environment produced a guidance note entitled 'Interpretation of Major Accident to the Environment for the Purposes of the CIMAH Regulations' which describes what constitutes a major accident to the environment in this context.[15] The guidance indicates the impact on a number of elements of the environment which would be considered to be a major accident in order to provide an indication to both industry and regulators as to the level of consequences which are relevant.

Regulation 4 of the CIMAH Regulations requires manufacturers to demonstrate that they have identified major accident hazards to people or the environment arising out of their activities and have taken adequate steps to prevent such major accidents and limit their consequences. The extent to which the application of this regulation has led to a proper risk-based consideration of major accidents to the environment has been variable, but the successor regulations under COMAH should provide more detailed and specific indication of what is required, leading to an improvement in the quality of environmental aspects of major accident safety reports.

COMAH (*Control of Major Accident Hazards*)

The so-called Seveso II Directive, COMAH, has a number of significant differences to CIMAH and must be implemented in the UK by February 1999. It is likely that the HSE and the Agency will be joint competent authorities for these regulations. In the second 'Seveso Directive', which was recently adopted, the need to apply a risk-based approach to protection of the environment from major accidents was more explicitly stated. At the time of writing this article, the regulations and associated guidance have yet to be produced, but a number of items of supporting R&D have been undertaken, to identify what is realistic with regard to application of a risk-based approach and develop model assessment techniques to assist industry in meeting its obligations.

In 1995 a R&D project was undertaken to produce a discussion document on 'Environmental Risk Criteria'.[16] This project was funded primarily by the European Union, with co-funding by UK regulators and a number of industries. The purpose of the project was to put forward, for consideration, an approach to applying risk assessment techniques to protection of the general environment. The report considered the aquatic environment alone and specifically looked at rivers and lakes. A key element of the approach put forward was the use of an

---

[15] Department of the Environment, Toxic Substances Division, *Interpretation of Major Accident to the Environment for the Purposes of the CIMAH Regulations—A Guidance Note by the Department of the Environment*, DoE, London, 1991.

[16] AEA Technology, VROM, AGEL-CBI, *Environmental Risk Criteria for Accidents: A Discussion Document*, NRA, Bristol, 1995.

Environmental Harm Index (EHI), based on a combination of the toxicity of the substance for the most sensitive member of the vulnerable ecosystem and the size of ecosystem which would be affected.

Following on from the 'Environmental Risk Criteria' project, a second item of work, which came to be known as the 'Management of Harm to the Environment' project, was undertaken. This project had the purpose of developing the concepts from the earlier approach into something which might be used to provide guidance to industry on how it might meet its environmental obligations under the COMAH directive, as well as developing environmental risk criteria for more general application. This second phase of the work introduced a third element to an EHI, to incorporate a factor representing the recovery time following an accident.

Industry has been involved in the debate about the EHI concept and the associated frequencies which have been proposed, and a number of valid concerns about the initial proposals have been voiced. The proposed methodology has been revised in light of these concerns and any guidance which is based on this work will stress the need to understand the uncertainties in any assessment. COMAH Regulations and guidance will be subject to consultation before they are implemented but, in order to comply with Annex III of the Directive, operators must have a safety management system which addresses 'adoption and implementation of procedures for systematically identifying major hazards arising from normal and abnormal operation and assessment of their likelihood and severity'. The adoption of this approach will have major benefits in protection of the environment at those sites which are covered by COMAH and the guidance which is produced may be transferable to non-COMAH sites in appropriate situations.

5 The Future

The Environment Agency is required, under the legislation which brought it into existence, to discharge its functions, contributing towards attaining the objective of achieving sustainable development, taking into account likely costs and benefits. Risk management is an ideal tool to help ensure achievement of these requirements, both strategically and in specific areas or locations. There is a clear link between risks, benefits and costs, although it is often difficult to quantify the value, in monetary terms, of protection of a particular element of the environment. However, the ability of risk management to target resources, following methodical consideration of potential impacts, is fundamental to cost-effective environmental protection, which is essential if a sustainable society is to be affordable.

There is often resistance to the application of formal risk management techniques because they are considered, by many, to be esoteric. However, we all undertake numerous informal risk assessments as part of our everyday lives. We may be prepared to take a risk where the frequency of an adverse event is relatively high but the consequence is small, and for each permutation of frequency and consequence we make decisions about acceptability. The attraction of a more formal approach to risk is greater consistency, standardization and a systematic approach. With regard to pollution risk management, the application of proportionate techniques is seen by the Environment Agency as a key means of ensuring that proper consideration is given to the damage which might be caused to the environment by any of man's ventures or activities, before deciding whether they should proceed, with or without conditions.

The greatest number of accidental releases of chemicals are made to the aquatic environment and this is where the greatest progress has been made in developing quantified risk assessment techniques. Over the past six or seven years, techniques have been developed which cover many of the key elements of aquatic pollution risk management, from looking at the likelihood of a chemical spill, to modelling concentrations in the environment and the frequency with which they might be anticipated. Consideration has also been given to the question of 'what is an acceptable risk?' Risk criteria are a difficult issue, as they embrace all aspects of a society's capability and willingness to pay to protect its environment. The criteria which have been put forward may not be definitive, but they provide an initial position which can be refined in the light of experience.

Because of the need to exploit fully the benefits of risk assessment for targeting of resources to protect and improve the environment, across its range of responsibilities, the Agency has established a National Centre for Risk Analysis and Option Appraisal, based in London. This centre provides a focus for expertise, both nationally and internationally. An important element of its work is to manage R&D so that the appropriate, cost-effective tools are available to ensure that the full benefits of a risk-based approach are available, to allow the best possible protection and improvement of the environment within the resources available. The centre will also manage the convergence of techniques inherited from its predecessor organizations, to ensure a coherent suite of techniques is available.

The Agency now produces Local Environment Agency Plans (LEAPs), in consultation with interested parties, to cover all of its area of responsibility, and these may provide an ideal vehicle for a risk-based approach to management of the whole environment.

In one way or another, risk-based methods will play an increasingly important part in the way that the Environment Agency achieves its vision for 'A better environment in England and Wales for present and future generations'.

Quantitative Cancer Risk Assessment—Pitfalls and Progress

STEVE E. HRUDEY

1 Introduction

Risk assessment, in a qualitative sense, has been a fundamental adaptation for human survival in a competitive ecosystem. Throughout history, the ability to recognize and respond to danger has allowed the human species to achieve unprecedented longevity. However, health risk assessment, in a quantitative sense, is a phenomenon which has emerged only over the past two decades. Demands emerged in the early 1970s for methods to quantify the risks posed by carcinogens in the environment. The nature of those demands shaped current practices in quantitative cancer risk assessment.

This article will focus primarily on experience from the United States because the practice of quantitative cancer risk assessment was born there and has become entrenched in their environmental regulatory system to a greater degree than in any other country. The risk assessment literature has been dominated by US risk assessment practices and issues. While other countries have developed their own cancer risk assessment policies to varying degrees,[1] they have been inevitably influenced by US experience.

Cancer has been recognized as an insidious and often fatal human disease at least since the 5th century BC, when Hippocrates likened to crabs the pattern of blood vessels emanating from a tumour.[2] He named the disease karkinoma in Greek, or cancer in Latin. Although cancer is commonly treated as if it was a single disease, medical science recognizes cancer to be a family of several diseases. These diseases share the common features of undisciplined growth of cells (tumours) which can invade surrounding tissues and spread to other sites in the body (metastasis). These characteristics, which often lead to a withering of the body and a slow, often painful death, have made cancer a highly dreaded disease. Taken together with the major prevalence of cancer, public concern about the

---

[1] J. Whysner and G. M. Williams, *Regul. Toxicol. Pharmacol.*, 1992, **15**, 41.

[2] S. Reiser, *Medicine and the Reign of Technology*, Cambridge University Press, Cambridge, 1978, pp. 77–79, as cited in J. T. Patterson, *The Dread Disease. Cancer and Modern American Culture*, Harvard University Press, Cambridge, MA, 1987, p. 12.

causes of cancer is certainly justified. Projecting current rates and averaging for both sexes, about 39% of Canadians will contract cancer during their lifetime and about 25% will die of cancer.[3]

Although quantitative cancer risk assessment did not develop until the mid-1970s, demands for some governmental response to growing cancer rates can be traced much earlier in the US. Unsuccessful attempts to commit federal funds towards cancer research by the National Academy of Sciences in 1928 were ultimately followed by the successful creation of the National Cancer Institute (NCI) in 1937. Even then, cancer was a serious concern for the public. Gallup polls asking 'What disease or illness would you dread having most?' found cancer named by 76% of respondents in 1939 and by 67% of respondents in 1947.[4]

The NCI created a new Environmental Cancer Section in 1948 headed by Wilhelm Hueper, a physician who had been fired from the chemical industry after disputes with management over his findings that β-naphthylamine was a cause of bladder cancer among dye manufacturing workers. Following this experience with irresponsible industrial managers, Hueper wrote a treatise on occupational tumours. Along with his work on identifying chemical carcinogens at the NCI, he was a strong influence on Rachel Carson who devoted an entire chapter of *Silent Spring*[5] to what she described as a growing cancer risk posed by man-made environmental pollution.

Silent Spring had an enormous influence on the emergence of a modern environmental ethic. The US Environmental Protection Agency (US EPA), created in response to the environmental concerns of the 1960s, was inevitably influenced by the prevailing perspectives on the threat of environmental carcinogens. The political importance of cancer was reflected in President Nixon's announcement of a 'War on Cancer' by means of the National Cancer Act of 1971.

The view that environmental contaminants were a major factor in human cancers had been reinforced by a World Health Organization report[6] in 1964 claiming that three quarters of all cancers were caused by extrinsic factors (those other than genetic predisposition). The proportion attributed to extrinsic factors was later extended to as high as 90% of all cancers.[7] Extrinsic factors, which included all factors operative on an individual after birth (diet, smoking, alcohol consumption, sexual behaviour), were also expressed as 'environmental factors' which some environmentalists readily equated with pollution by man-made chemicals.

Debate over the role of environment in cancer causation was highly charged. One popular book[8] promoting the environmental pollution theory of cancer described it as 'the plague of the twentieth century'. A 1975 article[9] titled 'The

[3] National Cancer Institute of Canada, *Canadian Cancer Statistics* 1997, Canadian Cancer Society, Toronto, 1997, p. 45.
[4] J.T. Patterson, *The Dread Disease. Cancer and Modern American Culture*, Harvard University Press, Cambridge, MA, 1987.
[5] R. Carson, *Silent Spring*, Fawcett World Library, New York, 1962, chap. 14, pp. 195–216.
[6] World Health Organization, *Prevention of Cancer*, Technical Report Series 276, WHO, Geneva, 1964.
[7] J. Higginson, in *Proceedings of the 8th Canadian Cancer Conference*, Pergamon Press, 1969, pp. 40–75.
[8] S.S. Epstein, *The Politics of Cancer*, Anchor Press/Doubleday, Garden City, NY, rev. edn., 1979.
[9] L. Ember, 'The specter of cancer'. *Environ. Sci. Technol.*, 1975, **9**, 1116.

Specter of Cancer' in a major environmental science journal stated that 'environmentally-caused diseases are definitely on the increase'. Cancer also provided the environmental movement with a compelling metaphor for the rampant postwar industrial growth and consumerism which was responsible for substantial environmental degradation. These perspectives raised high expectations for the health benefits of regulating environmental carcinogens.

2 Early History

Origins of Cancer Risk Assessment

The early agenda of the US EPA included the prohibition or restricted use of carcinogenic pesticides, an obvious preventable exposure. US EPA lawyers sought court adoption of a set of 'principles' on cancer as 'officially noted facts'.[10] These wide-ranging 'principles' uncritically mixed prevailing knowledge with supposition and regulatory policy. They drew scathing criticism from many scientific sources, including an editorial in a respected medical journal[11] which described the 'principles' as ranging from 'the innocuous to the absurd'. Many of these statements might have been reasonable proposals for a cautious policy of regulating carcinogens, but the attempts to enshrine them as scientific facts was justifiably criticized.

The critical questioning of the scientific rationale for these regulatory initiatives led to creation of committees to develop guidelines which would more accurately reflect the existing consensus of scientific knowledge about carcinogenic risk. They proposed that risk assessment should answer two main questions:[12]

(1) How likely is an agent to be a human carcinogen?
(2) If an agent was presumed to be a human carcinogen, what is the estimated impact on human health?

The first question requires mainly a qualitative evaluation while the second requires quantitative risk assessment.

A Carcinogen Assessment Group (CAG) was created to evaluate the evidence on the first question for any agent by considering the known and possible exposure patterns, metabolic characteristics, experimental animal studies and epidemiologic studies. Evaluating the qualitative issues depended on a number of premises comprising a weight of evidence approach.[12] This approach established a hierarchy of evidence in terms of their probative value for answering the first question.

Overall, the 'strongest' evidence for carcinogenicity in humans was judged to arise from adequate epidemiological data supported by relevant confirmatory evidence from animal studies. Because such data were not generally available for most environmental agents, reliance on animal test results became inevitable.

Given the difficulties recognized with basing the quantitative estimation of

---

[10] R. E. Albert, *Crit. Rev. Toxicol.*, 1994, **24**, 75.
[11] Anon., *Lancet*, 1976, March 13, 571.
[12] US EPA, *Fed. Regist.*, 1976, **41**, 21–402.

human cancer risk upon evidence from animal experiments and the need to extrapolate from high-level experimental exposures to low-level environmental exposures, the techniques for making such estimates were originally described as 'very crude' and the estimates they provided were characterized as only 'rough indications of effect'.[12] The answers to the second question were to be based upon an evaluation of exposure patterns and dose–response relationships to develop quantitative estimates of cancer risk using 'a variety of risk extrapolation models'. This assessment was supposed to indicate clearly the uncertainties in the data and the extrapolation technique, as well as offering some judgements about the relevance to human cancer risk of the experimental animal studies.

The original 1976 guidelines were generic and very brief, but they did reflect consideration of many of the contentious issues which have emerged in the subsequent practice of quantitative cancer risk assessment. The guidelines allowed for a wide range of interpretation, but the US EPA took the perspective that 'any evidence of tumorigenic activity in animals is a signal that the agent is a potential human carcinogen'.[13] This position was justified on the basis that an appropriate animal model had been demonstrated for all but one known human carcinogens at that time, indicating that false negatives were not a major problem in relying on animal results. The problem of false positives, concluding incorrectly that an animal carcinogen was in fact also a human carcinogen, was acknowledged as a possibility, but was only a limited concern because of the lack of evidence about how frequently or with what types of agents false positives might arise.

No Threshold Hypothesis of Carcinogenesis

One theme became the single most important aspect of quantitative risk assessment for carcinogens. This was the assumption that there was no threshold in the dose–response relationship for a carcinogen. The no-threshold premise was bound to become a central controversy because, unlike a conventional scientific hypothesis, this premise cannot be tested or falsified.[10] The premise was originally a science policy decision, made for the purposes of being cautious in the protection of public health. Despite any original limited intentions, this policy was often expressed in a more dogmatic manner. For example, an *ad hoc* committee of the Surgeon General had stated in a report for hearings before the US Senate that: 'No level of exposure to a chemical carcinogen should be considered toxicologically insignificant for man'.[14] Later, senior US EPA officials stated: 'Such a dose–response pattern implies that a safe level of exposure is nonexistent'.[13] This premise effectively renders 'unsafe' any guidelines for exposure to carcinogens, regardless of the evidentiary basis for them. This issue is considered further in Section 4.

An all-or-nothing concept of carcinogenesis, whereby any exposure to a

---

[13] R. E. Albert, R. E. Train and E. Anderson, *J. Natl. Cancer Inst.*, 1977, **58**, 1537.

[14] National Cancer Institute, Ad Hoc Committee, *Evaluation of Low Levels of Environmental Chemical Carcinogens*, Report to the Surgeon General, April 22, 1970, as cited in Office of Technology Assessment Task Force, *Identifying and Regulating Carcinogens*, Lewis Publishers, Chelsea, MI, 1988, p. 32.

carcinogen will result in cancer, has never been supported by animal experiments nor epidemiologic studies. Yet, a view that any exposure to a carcinogen may be dangerous has pervaded the history of carcinogen risk management. In 1992, 62% of a national sample of the Canadian public either agreed or strongly agreed with the statement:[15] 'If a person is exposed to a chemical that can cause cancer, then that person will probably get cancer some day'.

Much of the early focus on controlling carcinogens was aimed at the total elimination of exposure. This was reflected in the 1958 enactment of the Delaney clause by the US Food and Drug Administration (FDA), prohibiting the use of any food additives which had been determined to cause cancer. Such prohibitions were at least conceptually feasible for substances which were intentionally added to food, but they posed a dilemma for substances which may be present as unintentional residues.

In most cases, analytical detection limits for trace substances were poor enough that inability to detect a substance was common and could be casually treated as complete absence. However, during the 1960s analytical chemistry began to achieve major advances in sensitivity which have continued to accelerate and make obsolete any policy for complete absence of carcinogens. As an illustration, the current practical detection limits for routine laboratory analysis of drinking water for benzene, a human carcinogen, is about $0.1~\mu g\,L^{-1}$. Benzene below this level would be reported as non-detected. Detectability has improved by over 10 000-fold since the 1960s, now finding benzene in waters previously considered benzene-free. Yet, $0.1~\mu g\,L^{-1}$ corresponds to 8×10^{14} molecules of benzene per litre of water. This example illustrates the enormous scope for future lowering of detection limits to enable finding carcinogens in waters previously assumed to be carcinogen-free. Clearly, a zero tolerance policy relying on analytical detection limits is arbitrary and unrelated to health risk.

Early Models of Carcinogenesis

The foregoing developments raised the need for some means to judge the safety of non-zero levels of exposure. The first rigorous attempt at proposing an acceptable non-zero level of exposure to a carcinogen was put forth in 1961.[16] Mantel and Bryan elaborated the practical limitations of animal bioassay experiments for defining safe levels by noting that a dose level which yielded zero excess tumours in 100 test animals still corresponded to a risk as high as 4.6 in a 100, at a 99% confidence level. This minimum detectable risk level is higher than 1 in 100 because of the statistical need to account for sampling variability in the test population at the specified confidence level. By the same logic, it would be necessary to find no tumours in 460 million test animals to be confident at a 99% level that the probability of a tumour occurring was less than 1 in 100 million. Although the rationale for choosing 1 in 100 million as a measure of 'virtual' safety was not explained, it may have been that this level would have corresponded to only about 1 case of cancer in the entire US population of the 1960s.

Because the practical limitations of experimental bioassays preclude direct

[15] D. Krewski, P. Slovic, S. Bartlett, J. Flynn and C. K. Mertz, *Human Ecolog. Risk Assess.*, 1995, **1**, 231.
[16] N. Mantel, and W. R. Bryan, *J. Natl. Cancer Inst.*, 1961, **27**, 455.

measurement of low risk levels, Mantel and Bryan proposed estimating the carcinogen dose for an upper bound risk of 1 in a 100 million by noting that experimental results for quantal (tumour or no tumour) results often fit a log normal probability distribution (log for dose, probability for tumour response). They argued that the shallowest (and therefore most cautious) slope they would expect to observe, based on their extensive experience with animal bioassays, would be 1 normal deviation per log (factor of ten) of dose. Accordingly, they calculated that a dose which produced zero tumours in 50 animals should be divided by 18 000 to have 99% confidence that the corresponding risk would be less than 1 in 100 million.

This pioneering attempt at estimating risk levels from animal bioassays was useful as an alternative to the unworkable zero concept. However, it was predicated on acceptance of the log–probit model for dose–response, or at least acceptance that using an extrapolation slope of 1 probit per log was sufficiently cautious. Ironically, this approach was abandoned because it could not be defended on the basis of any mechanistic evidence.

Earlier quantitative models of carcinogenesis had proposed relationships between dose and cancer response within the range of observable data.[17–19] Armitage and Doll[20–22] observed that human cancer death rates were consistent with a multistage process (six or seven stages), from which they developed their multistage model. This accounted for the long latency period between exposure to a carcinogen and the appearance of cancer as well as a linear relationship between cancer incidence at a given age and the dose of the initial carcinogen.[21]

As originally presented, these models did not tackle the problem addressed by Mantel and Bryan,[16] that of extrapolating from high experimental animal doses to the lower doses observed in human experience. Simultaneously, work on radiation exposure considering leukaemia risk for Hiroshima and Nagasaki atomic bomb survivors and for ankylosing spondylitis patients receiving X-ray treatments suggested that cancer risk could be extrapolated linearly with no apparent threshold.[23] The concept of extrapolating the dose–response curve linearly through zero dose was subsequently adopted for the regulation of cancer risk from radiation exposures.[24] This pragmatic policy choice has since often been treated as scientific fact.

[17] S. Iverson and N. Arley, *Acta Pathol. Microbiol. Scand.*, 1950, **27**, 773.

[18] J.C. Fisher and J.H. Hollomon, *Cancer*, 1951, **4**, 916.

[19] C.O. Nordling, *Br. J. Cancer*, 1953, **7**, 68.

[20] P. Armitage and R. Doll, *Br. J. Cancer*, 1954, **8**, 1.

[21] P. Armitage and R. Doll, *Br. J. Cancer*, 1954, **8**, 1.

[22] P. Armitage and R. Doll, in *Proceedings of the 4th Berkeley Symposium on Mathematical Statistics and Probability*, University of California Press, Berkeley, CA, 1961, vol. 4, pp. 19–38.

[23] E.B. Lewis, *Science*, 1957, **125**, 965.

[24] BEIR, in *Report of the Advisory Committee of the Biological Effects of Ionizing Radiations*, National Academy of Sciences, National Research Council, US Government Printing Office, Washington, DC, 1972.

3 Regulatory Practice

Basic Framework

The early history of cancer risk assessment demonstrates the controversial climate in which it was born. Despite the efforts of various regulatory agencies to seek wider scientific support, controversies surrounding management decisions relying on risk assessments continued. In a further attempt to achieve scientific respectability, a report to guide regulatory risk assessment was commissioned from the National Research Council of the US National Academy of Sciences.[25] This publication, widely referred to as the Red Book became extremely influential in guiding risk assessment practice during the 1980s and early 1990s.

The Red Book codified the emerging consensus that risk assessment consisted of four steps: hazard identification, dose–response assessment, exposure assessment and risk characterization. Hazard identification was a largely qualitative step aimed at evaluating the weight of evidence from human epidemiology studies and experimental animal bioassays to determine the likelihood of a chemical being a human carcinogen. This addressed the first qualitative risk question posed in the 1976 guidelines, while the remaining three steps addressed the second, quantitative risk question.

Dose–response assessment evaluated the quantitative evidence from animal bioassays or, less commonly, epidemiologic studies to estimate cancer risk as a function of carcinogen exposure. Because environmental exposures are almost invariably much lower than the dosage ranges in animal experiments or in occupational epidemiology studies, extrapolation models had to be adopted to estimate the cancer risks associated with environmental exposures.

Exposure assessment evaluated the character and level of exposure to carcinogens which occurs in the population under consideration. The character of exposure includes the specific chemical forms, exposure routes (inhalation, ingestion of food and water or dermal contact) and the time course of exposure (single, intermittent or continuous). Risk characterization involved a combination of the findings from the exposure assessment and the dose–response assessment to estimate the incidence of the adverse effect in the exposed public. This stage of risk assessment was also charged with elaborating the uncertainties in the risk estimates.

Although the Red Book is most commonly cited for its recommendation to separate risk assessment from the political, social and economic imperatives which drive risk management decisions, it provided an extensive and thoughtful collection of the range of science policy choices which are inherent in risk assessment, listing 49 issues and associated questions. The failure to distinguish science policy choices from scientific 'facts' was the major criticism of the earlier cancer 'principles'. Much of the re-occurring misunderstanding in the practice of risk assessments and subsequent public disputes over their predictions can be traced to a continuing failure to recognize the overwhelming role played by these science policy choices.

---

[25] National Research Council, *Risk Assessment in the Federal Government: Managing the Process*, National Academy Press, Washington, DC, 1983.

Table 1 Major default assumptions* in cancer risk assessment guidelines (Reproduced with permission from ref. 27)

1. Laboratory animals are a surrogate for humans in assessing cancer risks; positive cancer-bioassay results in laboratory animals are taken as evidence of a chemical's cancer-causing potential in humans
2. Humans are as sensitive as the most sensitive animal species, strain or sex evaluated in a bioassay with appropriate study design characteristics
3. Agents that are positive in long-term animal experiments and also show evidence of promoting or co-carcinogenic activity should be considered as complete carcinogens
4. Benign tumours are surrogates for malignant tumours, so benign and malignant tumours are added in evaluating whether a chemical is carcinogenic and in assessing its potency
5. Chemicals act like radiation at low exposures (doses) in inducing cancer, *i.e.* intake of even one molecule of a chemical has an associated probability for cancer induction that can be calculated, so the appropriate model for relating exposure–response relationships is the linearized multistage model
6. Important biological parameters, including the rate of metabolism of chemicals, in humans and laboratory animals, are related to body surface area. When extrapolating metabolic data from laboratory animals to humans, one may use the relationship of surface area in the test species to that in humans in modifying the laboratory animal data
7. A given intake of a chemical has the same effect, regardless of the time of its intake; chemical intake is integrated over time, irrespective of intake rate and duration
8. Individual chemicals act independently of other chemicals in inducing cancer when multiple chemicals are taken into the body; when assessing the risks associated with exposures to mixtures of chemicals, one treats the risks additively

*The above are default assumptions adopted for the pragmatic purpose of implementing cancer risk assessment; they are not scientific facts.

Guiding Policy and Default Assumptions

Despite early recognition of the need for making clear distinctions between science policy and scientific evidence, these have remained substantially confounded in risk assessment practice. The inferential extrapolations from evidence to conclusions arising from many of the assumptions simply cannot be verified. Consequently, risk assessment practices which take these assumptions as valid have been adopted as matters of policy, generally guided by an overall preference for 'conservatism'. In this context, conservative assumptions mean those which are more likely to overestimate rather than underestimate risk, thereby erring on the side of caution.

The policy of conservatism was largely instituted by means of default assumptions. Those which were embodied in the US EPA 1986 Cancer Risk Assessment Guidelines[26] have been identified[27] and major default assumptions are summarized in Table 1. The 1986 guidelines[26] have been the major reference

[26] US EPA, *Fed. Regist.*, 1986, **51**, 33–992.
[27] National Research Council, *Science and Judgement in Risk Assessment*, National Academy Press, Washington, DC, 1994.

source for regulatory practice in quantitative cancer risk assessment over the past decade.

Hazard Identification.   This stage was to provide a review of information about the agent concerning its physical chemical properties, environmental pathways, patterns and intake routes of human exposure, structure–activity relationships, metabolic and pharmacokinetic properties, evidence of toxicological effects other than carcinogenicity, short-term tests for genetic damage, and results from long-term animal studies.[26] In practice, animal study results have been the dominant influence on quantitative cancer risk assessment. Positive tumour results from animal bioassays have been treated, as a matter of policy, as an indication that the agent is potentially a human carcinogen. The degree of that potential risk has been treated as more credible if the response observed in exposed animals compared to the unexposed control animals:

- Is strong in the sense of a large number of tumours which occur with a high level of statistical significance
- Shows a consistent dose–response relationship either in overall tumour occurrence or in shortening of time to tumour occurrence or time to death with tumours
- Occurs in both sexes
- Occurs in more than one experimental species or strain
- Occurs in different experiments
- Is evident at more than one site
- Is evident as more malignant rather than benign tumours as a function of dose

While factors like these have a substantial bearing on determining the weight of evidence for the qualitative answer to the question about the likelihood of a hazard being a human carcinogen, there is no 'correct' way to translate such factors into the quantitative results used for dose–response modelling.

Policy has also dictated, on grounds of conservatism, that the most sensitive animal responses were to be used for estimating the human response. While this is clearly the cautious route for quantitative risk estimation, it raises a question of relevance if the most sensitive rodent response is not one which is physiologically relevant to humans.

The long-term animal studies were performed with dose levels at or near the maximum tolerated dose (MTD) in order to maximize the sensitivity of the experiment. For these purposes, MTD is defined as the highest dose 'which when given for the duration of the chronic study, is just high enough to elicit signs of minimal toxicity without significantly altering the animal's normal life span due to effects other than carcinogenicity'.[28] This dose, which is normally estimated from a 90 day, sub-chronic study, should cause less than a 10% decrement in weight gain and it should not comprise so large a portion of the normal diet as to lead to nutritional imbalance. Two additional doses were recommended, typically 1/2 and 1/4 of MTD. This practice raised opposing concerns which were

---

[28] Office of Science and Technology Policy, *Fed. Regist.*, 1985, **50**, 10372.

mentioned in the guidelines.[26] On the one hand, such high doses might invoke physiological and cellular level responses (*e.g.* overwhelming of detoxification or elimination mechanisms), leading to tumour formation which would not occur at much lower dosage levels. On the other hand, for carcinogens which must be metabolically activated, high doses might overwhelm the metabolic activation capacity, resulting in an underestimation of carcinogenic activity at high dose.

Other controversial issues in hazard identification were acknowledged in the 1986 guidelines.[26] These included the possibility of agents being promoters rather than tumour initiators and the significance to human health risk for certain classes of rodent tumours, such as mouse liver tumours. These controversies are discussed further in Section 4.

Although human studies were recognized as potentially the most relevant evidence for estimating human cancer risk, their limited availability and capability, particularly for detecting risk at relevant exposure levels, led to little discussion in the 1986 guidelines.[26] The limited utility of descriptive studies for the purposes of supporting causal inference was noted, with emphasis being primarily placed on analytical study designs like case-control and cohort studies.

The main product of the hazard identification process was the classification of a given carcinogen. The US EPA regulatory scheme was an adaptation of the classification originally developed by the International Agency for Research on Cancer (IARC). The scheme considered primarily the human epidemiologic and long-term animal study evidence to place an agent into one of categories A through E, as outlined in Table 2. These categories were intended to summarize deliberations on the weight of evidence for answering the first question, concerning how likely the agent is to be a human carcinogen. Substances classified into groups A and B were deemed suitable for further analysis by quantitative risk assessment. Even those in group C were generally deemed suitable, but individual judgements could be made to exclude such substances from quantitative analysis, along with those in groups D and E.

Dose–Response Assessment.   Simple linear extrapolation with no threshold was adopted for cancer risk assessment, with several arguments commonly invoked as justification.[29] First, somatic mutation theory held that cancer is the result of an initiating event between a single molecule of a carcinogen and a DNA molecule, resulting in a mutation leading to a line of damaged cells which ultimately develop into a tumour. According to the theory, without considering any other factors such as the active carcinogen having to reach the DNA, the mutation having to occur at the correct gene, the cell having to survive to replicate and the damaged DNA having to avoid being repaired, the probability or chance of cancer occurring will be directly proportional to the number of carcinogen molecules.

Second, the analogy with radiation was invoked to suggest that because some carcinogens can damage DNA, as does radiation, the apparent low dose linearity for radiation risk could be applied to chemical carcinogens. Critics of this

[29] International Life Sciences Institute, *Low-Dose Extrapolation of Cancer Risks—Issues and Perspectives*, ed. S. Olin, W. Farland, C. Park, L. Rhomberg, R. Scheuplein, T. Starr and J. Wilson, ILSI Press, Washington, DC, 1995.

Human evidence	Animal evidence				
	Sufficient	Limited	Inadequate	No data	No evidence
Sufficient	A	A	A	A	A
Limited	B1	B1	B1	B1	B1
Inadequate	B2	C	D	D	D
No data	B2	C	D	D	E
No evidence	B2	C	D	D	E

Table 2 US EPA classification* of carcinogens according to 1986 Guidelines (Reproduced with permission from ref. 26)

*These assignments are presented for illustrative purposes. There may be nuances in the classification of both animal and human data indicating that different categorizations than those given in the table should be assigned. Furthermore, these assignments are tentative and may be modified by ancillary evidence. In this regard, all relevant information should be evaluated to determine whether the designation of the overall weight of evidence needs to be modified. Relevant factors to be included along with tumour data from human and animal studies include structure–activity relationships, short-term test findings, results of appropriate physiological, biochemical and toxicological observations and comparative metabolism and pharmacokinetic studies. The nature of these findings may cause an adjustment of the overall categorization of the weight of evidence.

Group A   Human carcinogen—sufficient evidence from epidemiologic studies to support a causal association between exposure to the agent and cancer.

Group B   Probable human carcinogen—limited evidence from epidemiologic studies (B1) or sufficient evidence from animal studies, and no data or no evidence from epidemiologic studies (B2).

Group C   Possible human carcinogen—limited evidence of carcinogenicity in animals in the absence of human data.

Group D   Not classifiable as to human carcinogenicity—inadequate human and animal evidence of carcinogenicity or for which no data are available.

Group E   Evidence of non-carcinogenicity for humans—negative evidence in at least two adequate animal tests in different species or in both adequate epidemiologic and animal studies.

argument note that radiation is not altered by metabolic processes nor does its arrival at a target site normally depend upon non-linear physiological processes.

Third, additivity to background dose holds that every individual experiences some level of exposure to chemical carcinogens from a wide variety of background sources and that any estimated regulatory exposure will be additional to this background level. Provided that there is no threshold, it can be shown mathematically that, for a range of dose–response curves, small incremental additions of the carcinogen will increase the risk in an approximately linear fashion.[30] Likewise, an argument can be made that carcinogens which act *via* mechanisms identical to those responsible for background cancer rates will increase risk in a linear fashion by incrementally adding to the background rate.

Finally, given the public health policy preference for using an upper bound on estimates of cancer risk, it has been demonstrated that the upper bound estimates are approximately linear at low dose and relatively insensitive to model choice.[31] Consequently, if the policy choice is to use upper bound risk estimates, this argument holds that low dose linearity will follow.

[30] K. S. Crump, D. G. Hoel, C. H. Langley and R. Peto, *Cancer Res.*, 1976, **36**, 2973.
[31] H. Guess, K. Crump and R. Peto, *Cancer Res.*, 1977, **37**, 3475.

Initially these arguments for a no-threshold linear extrapolation were implemented by means of the one-hit model which expressed the probability of a tumour as a straight linear function of dose with a zero intercept. Although other models were proposed, in 1980 the US EPA adopted the linearized multistage model, LMS,[32] as their default model to replace the one-hit model.[33]

The LMS model expresses the lifetime probability of cancer, $P(d)$ at dose d as:

$$P(d) = 1 - \exp[-(q_0 + q_1 d + q_2 d^2 + \ldots q_k d^k)]$$ (1)

Additional risk above background for dose d, $A(d)$, is expressed as:

$$A(d) = [P(d) - P(0)]/[1 - P(0)]$$ (2)

The conservative practice in using the LMS model for regulatory purposes involved estimating upper 95% confidence values for the coefficients 'q' which were used to fit the animal bioassay data. These upper bound estimates were designated q^* by convention. Consequently, the equation for risk at low doses, where the dose terms taken to higher powers will drop out, leaves:

$$P(d) = 1 - \exp[-(q_0^* + q_1^* d)]$$ (3)

so that background risk at $d = 0$ becomes

$$P(0) = 1 - \exp(-q_0^*)$$ (4)

making the extra risk $A(d)$

$$A(d) = [P(d) - P(0)]/[1 - P(0)] = 1 - \exp(-q_1^* d)$$ (5)

Furthermore, at very small doses (d), the mathematics reduce to

$$\exp(-q_1^* d) \approx 1 - q_1^* d$$ (6)

Consequently, the LMS expression for extra risk at low dose becomes approximately:

$$A(d) = q_1^* d$$ (7)

This LMS expression for low dose exposures is similar in form to the one-hit model in the sense of simply expressing excess risk (A) as a linear carcinogenic slope factor (q_1^*) times the dose (d). However, the LMS value of the slope term (q_1^*) differs numerically from the slope derived by the one-hit model because it is based upon an upper confidence bound for fitting of all of the bioassay response data rather than using an upper confidence estimate on only the lowest dose yielding a tumour response. Consequently, the US EPA adopted the LMS model

[32] K. S. Crump, *J. Environ. Pathol. Toxicol. Oncol.*, 1984, **5**, 339.
[33] E. L. Anderson, and the Carcinogen Assessment Group of the US Environmental Protection Agency, *Risk Anal.*, 1983, **3**, 277.

as their default extrapolation model on the grounds that it made fuller use of the bioassay data. Comparisons of slope factors between the two models differed only for cases of steeply rising dose–response curves, where the LMS slope could be lower than the one-hit estimate by less than five-fold.[33]

Scaling the dose responsible for observed levels of tumours between rodents and humans presents another choice affecting the quantitative predictions. The most commonly considered choices were scaling dose as either mass of agent per body mass per day ($mg\,kg^{-1}\,d^{-1}$) or mass of agent per body surface area per day ($mg\,m^{-2}\,d^{-1}$). Using the body mass basis results in approximately a six-fold lower human risk estimate when extrapolating from rat data and approximately 14-fold lower extrapolating from mouse data. Accordingly, the US EPA opted for the more cautious surface area basis for extrapolation which was approximated as body mass to the 2/3 power, assuming the surface area to volume relationship of a sphere. This makes the human equivalent dose equal to:[33]

$$d = [l_e m]/[L_e W^{2/3}] \tag{8}$$

where d = dose in $mg\,kg^{-1}\,d^{-1}$; l_e = duration of exposure (time units); m = average mass of contaminant dosed ($mg\,d^{-1}$); L_e = duration of experiment (time units); and W = body mass of the experimental animal (kg).

Data from epidemiologic studies, where judged to be adequate, were fitted with the best curve within the observed range and then extrapolated using a linear, no-threshold assumption to develop a carcinogenic dose–response slope factor for the excess risk.[33]

Although the guidelines adopted the LMS model as the default model, in the absence of adequate epidemiologic data to estimate dose–response, some important qualifiers were specified which were often overlooked in subsequent applications of the methodology. In particular, the guidelines stated[26] 'Irrespective of the options chosen, the degree of precision and accuracy in the numerical risk estimates currently do not permit more than one significant figure to be presented'.

Exposure Assessment. In theory, exposure assessment provides an opportunity to gather directly relevant evidence to estimate what levels of exposure to a carcinogen any specified human population has received. In practice, the exposure assessment is very much dependent upon specifying the purpose of the risk assessment which will specify the population to be considered. Once the population under study (hypothetical or real) has been determined, practical issues of characterizing the heterogeneity (variability) of exposure inevitably arise. Until the past few years, this variability was not fully addressed in an explicit fashion and one or more point estimates of exposure were used to represent the population.

In the case of the Superfund risk assessment programme, risk assessors were guided to estimate all possible carcinogen exposures *via* ingestion, inhalation and dermal contact to ensure that a complete estimation of dose had been achieved.[34]

[34] US EPA, *Risk Assessment Guidance for Superfund. Volume 1, Human Health Evaluation Manual, Interim Final*, Office of Solid Waste and Emergency Response, US Environmental Protection Agency, Washington, DC, 1989.

The reasonable maximum exposure (RME) was to be estimated for each pathway by using estimates of the upper 95% bound for each parameter input to the dose equation (*e.g.* concentration of carcinogen, ingestion or inhalation rate, exposure time, *etc.*).

In the case of the hazardous air pollutants programme, risk assessors were guided to estimate total population exposure as well as a maximally exposed individual (MEI). The MEI was supposed to represent an upper bound exposure in the population. This was implemented by assuming a hypothetical individual who would be located at the ground level location with the highest annual average air pollutant concentration as identified by a source dispersion model. Furthermore, the MEI was assumed to reside at that location 24 hours per day for 70 years and indoor pollutant concentrations were assumed equal to the outdoor concentration.[27]

Another default position was necessary for apportioning doses received for periods less than a lifetime. This was resolved by assuming linear proportionality such that a short-time high dose was treated for risk calculation purposes as if the same dose was prorated over an entire lifetime. This pragmatic decision assumes that a short-term intense dose would behave equivalently for developing a cancer risk as a long-term consistent dose over an entire lifetime. The lack of reality inherent in this approach was acknowledged in the 1986 guidelines,[26] but no other justifiable approach for dealing with this problem was apparent.

Risk Characterization. This stage was presented in two parts: a presentation of numerical estimates of risk and a framework for interpreting the significance of the risk.[26] The choices for numerical risk estimates were presented as using one or more of: a unit risk, a dose corresponding to a specified risk or the hypothetical individual or population risks.

The unit risk approach depends upon the linearity assumption and presents the findings in terms of the excess lifetime cancer risk associated with continuous lifetime (70 year) exposure to a given medium concentration (air, water, food). This was expressed either as the upper bound lifetime cancer risk of lifetime exposure to 1 unit of the medium concentration (μg L^{-1} for water, μg m^{-3} for air, μg g^{-1} for food) or as the medium concentration corresponding to a specified risk level (10^{-5}, 10^{-6} or 10^{-7}).

The dose corresponding to a specified level of risk need not depend upon using a linearized extrapolation model, but this was the usual practice. This dose, which was once referred to as a virtually safe dose (VSD), with reference to a one in a million lifetime cancer risk, is now commonly referred to as a risk specific dose (RSD) for any specified risk level.

The individual lifetime cancer risk may be expressed for a hypothetical individual with a specified average daily dose. Alternatively, a hypothetical population exposed to an average daily dose may be used to generate an estimate of the upper bound number of cancer cases possible in that exposed population over a specified duration.

In all cases, the standard protocol was intended to provide an upper bound risk estimate for hypothetical individuals or populations. The protocol was recognized as being incapable of determining the 'expected' risk for any real individual or population. The original guidelines attempted to caution against misinterpretation by stating that the LMS method producing an estimate of q_1* for subsequent calculations 'leads to a plausible upper limit to the risk that is consistent with some proposed mechanisms of carcinogenesis. Such an estimate, however, does not necessarily give a realistic prediction of the risk. The true value of the risk is unknown, and may be as low as zero'.[26]

The guidelines also recognized misunderstandings which might arise if the numerical risk estimates were decoupled from the qualitative, weight-of-evidence ranking. They called for always presenting the ranking together with the numerical risk estimate. However, risk assessment practice spread far beyond the US EPA to state agencies, consultants, industry and environmental advocacy groups and these cautions were often ignored. Risk estimates have been commonly expressed to several significant figures, adequate cautions about the upper bound nature of the risk estimate have been neglected and qualitative discussion of the weight-of-evidence with respect to human cancer has been presented separately from numerical estimates, if at all. Finally, and perhaps most serious, the hypothetical and generic nature of the risk calculation has not been communicated so that individuals have been left to ponder whether they or one of their family may be the unlucky one in a million.

Underlying Global Assumptions. The combination of major assumptions inherent in cancer risk assessment practice amounted to accepting that:[29]

- Dose–response data from limited size groups of laboratory experimental animals, of homogenous genetic make-up, can predict the dose–response relationship for a very heterogeneous human population
- Dose–response data obtained at high levels of exposure relative to expected environmental exposures can predict the dose–response relationship at these lower doses which are usually well below the experimental range
- Dose–response data obtained from consistent experimental dosing regimens can predict the dose–response relationships for much more variable human exposures

Applications

The quantitative risk assessment approach for carcinogens was applied in five major areas by the US EPA:[33] setting priorities, reviewing remaining risk after application of best available control technology, balancing risks with benefits, setting standards and target levels of risk, and estimating risks to specific populations such as residents near hazardous waste sites.

A major application for setting priorities was for ranking hazardous air pollutants. In this case, an index of potency for carcinogenic agents was developed in the form of unit risk factors. These values were estimates of lifetime cancer risk for a 70 kg human breathing a unit air pollutant concentration of $1 \mu g \, m^{-3}$ constantly over a 70 year lifetime. The resulting pollutant rankings were useful in the sense that they showed values ranging over a million-fold and they showed that air pollutants which had the strongest evidence of being human carcinogens (benzene, vinyl chloride) were also among the least potent pollutants by this measure.

In some cases where decisions needed to be made about the utility of further regulation to go beyond the best currently available control technology, quantitative risk assessment was used to consider the upper bound risk estimates associated with the emissions from that technology. Using this approach, the US EPA was able to conclude that additional emission controls for vinyl chloride would not be warranted because the upper bound cancer risks associated with application of best technology was already very low (*i.e.* less than 0.1 cases of cancer per year in the US).

In cases involving pesticide regulation, balancing benefits of the agent being reviewed for registration were routinely compared with upper bound estimates of cancer risk to determine the value of registering a pesticide. In situations involving pesticides of major agricultural importance or those which lacked registered alternatives, these factors would be used to justify tolerating exposure to a low cancer risk.

Review of upper bound risk estimates was also commonly used for environmental quality standards, notably water quality guidelines which generally posed upper bound, lifetime cancer risks within the range from 10^{-5} to 10^{-7}. In a few cases, such as disinfection by-products in drinking water, lifetime risks higher than 10^{-5} were deemed justifiable because of the obvious health benefits of disinfection of drinking water. In other cases, lifetime cancer risks lower than 10^{-5} led to risk reduction initiatives because there was limited benefit identifiable with the exposure.

Risk estimates also came to be applied to site-specific situations as the Superfund programme for identification and clean-up of hazardous waste sites was implemented during the 1980s. Many of these cases were adversarial in nature because of the legal actions involved in assigning financial responsibility for clean-up to parties whose waste had been disposed. Furthermore, the site-specific focus of risk assessment in these cases, and in cases where these techniques were applied for environmental health impact assessments of contentious industrial facilities like incinerators, brought the risk estimates to a more personal level. In these cases, the methods which may have been justifiable for estimating cancer risks in generic situations with hypothetical individuals at risk were being interpreted by real people in terms of their own personal risk. This led to inevitable miscommunication of risk because the hypothetical upper bound population risk estimates provided were inherently incompatible with the personalized, individual risk prediction which the publics in these circumstances were seeking.[35] In these situations, the utility of quantitative risk assessment is extremely limited compared with its enormous potential for creating misunderstanding.

4 Major Criticisms and Controversies

The practice of quantitative cancer risk assessment was born as a regulatory measure in response to the intense controversies associated with environmental

[35] V. R. Walker, *Risk Anal.*, 1995, **15**, 603.

chemical contaminants as causal factors in human cancer. The practical demands for answers greatly exceeded the capability of the tools and scientific knowledge base that was available. Because of the societal pressures for action, pragmatic assumptions were made to bridge the many gaps in knowledge, and answers were generated. Although the rationale for many of the assumptions was described and cautious qualifiers were specified, these were often either not communicated to or not understood by decision makers and affected publics. As a result, the predictions of cancer risk assessment were often used in ways for which they were not suited, leading to intense frustration in the community and growing criticism from the scientific establishment.[36,37] The malleability of the underlying assumptions for yielding different quantitative answers was exploited by environmentalists and industrialists alike to support their respective positions about the safety or danger of contaminants. An EPA administrator once likened risk assessment to a political prisoner: if you torture it long enough it will give you any answer you want. Overall, the credibility of cancer risk assessment has been severely damaged and the value of the process for guiding risk management decisions has suffered. Yet, no truly independent alternative approach has emerged, so the best prospects for future management of carcinogenic risks lie in recognizing the failings of past practice and adjusting future practice to minimize the recognized errors.

Expected vs. Upper Bound Cancer Risk

Perhaps the greatest single misunderstanding has been the belief that the risk assessment methods are capable of generating estimates of expected numbers of cancer cases rather than providing plausible upper bound estimates. Some may regard this distinction as subtle, but its importance to the credibility of the process is profound. In the absence of knowledge about specific mechanisms of cancer causation and means for detection by observation or experiment, educated choices of assumptions which are more likely to overestimate than to underestimate risk may still be defensible. However, when individual input values may vary in their magnitude by several factors of ten, an educated choice of the most likely value normally cannot be defended with the same confidence.

A recent survey[38] of environmental engineering professors and environmental epidemiologists indicated that 28% and 25%, respectively, either agreed or strongly agreed with the statement: 'A lifetime cancer risk as low as one chance in a million can be known for a given level of exposure to a carcinogen'. Although a majority of both disciplinary groups disagreed and a large proportion indicated that they did not know (17% for environmental professors, 27% for epidemiologists), the responses suggest that a substantial number of environmental specialists believe that risk assessment can provide accurate cancer risk estimates down to insignificant levels.

Some critics of cancer risk assessment have proposed compounding this

[36] P. H. Abelson, *Science*, 1994, **265**, 1507.
[37] P. H. Abelson, *Science*, 1995, **270**, 215.
[38] S. Rizak, C. G. Jardine and S. E. Hrudey, *Expert Judgements of Environmental Risks*, Eco-Research Report No. 97–2, University of Alberta, Edmonton, 1997.

S. E. Hrudey

misunderstanding by using maximum likelihood estimates (MLE) from dose–response modelling rather than the upper 95% confidence limits. Although the MLE is a better representation of the dose–response model curve fit within the region of the bioassay results, the meaning of MLE for the extrapolated linear slope is not clear.[29] An expectation that using MLE will produce an unbiased, best estimate for the linear slope will not be satisfied because the linear no-threshold models are strictly empirical, are not based upon mechanisms of cancer causation, and do not use independently measured mechanistic parameters. Consequently, their extrapolation to well beyond the experimental data range increases the parameter uncertainty too much for MLE to hold the meaning it does for the curve fitting of results within the observed experimental range. Differences between upper bound estimates and MLE linear slope estimates can range from two- to three-fold for roughly linear bioassay data up to 1000-fold in extreme, non-linear cases. The upper bound estimate is also much more stable to small differences in tumour count than the MLE.[29] The issue of extrapolated dose–response slope estimation is only one, albeit a very important, limitation of quantitative risk estimation, so the prospects of being able to replace upper bound cancer risk estimates with meaningful estimates of expected cancer risk at low environmental doses is simply not realistic for the foreseeable future.

In summary, quantitative cancer risk assessment has utility for implementing a cautious public health policy by estimating plausible upper bound cancer risks. These methods can also provide useful 'one way' assessments in cases where the upper bound risk estimates on a situation turn out to be too small to justify further risk management attention. However, the expectation that risk assessments can generate the equivalent of actuarial estimates of cancer occurrence will not be satisfied with any current or foreseeable quantitative cancer risk assessment methodology for lower dose exposures.

Mechanisms of Carcinogenesis

The single agent, chronic rodent bioassay has been the major evidence used for dose–response quantitation. This experiment is only able to determine whether an excess of tumours occurs in the exposed animals in relation to the control animals. A differential tumour response does not indicate anything substantive about mechanisms of carcinogenesis other than what can be deduced from the location and pathology of the tumour. The resulting tumour count which is used in the quantitative dose–response stage has not distinguished the possibility that excess tumours in the exposed *versus* control populations may have been caused by a mechanism which has a threshold.

The default practice in quantitative cancer risk assessment has been to ignore the distinction between initiation and promotion mechanisms unless exceptional evidence could justify an alternate approach. Experiments on the 139 chemicals (out of 222) which had tested positive by chronic bioassay for the US NCI/National Toxicology Program showed that 42% tested negative on the Ames salmonella-microsome assay and were deemed to be non-mutagenic.[39] Because mutagenesis

[39] J. Ashby and R. W. Tennant, *Mutat. Res.*, 1988, **204**, 17.

is the proposed mechanism for tumour initiation, these findings suggested the need for closer consideration of the tumour mechanism at work. The distinction of initiation from promotion is critically important to the subsequent dose–response modelling because a no-threshold model is a reasonably cautious policy which has only been rationalized for initiator (genotoxic, DNA-reactive) carcinogens. Promotion mechanisms are more likely to demonstrate a threshold, particularly if no reaction with a cell receptor is involved.[40]

Subsequently, a more comprehensive analysis of 522 rodent carcinogens and 55 human carcinogen exposures led to the classification of the rodent carcinogens into presumed DNA-reactive carcinogens and others. Using a structural alert classification for DNA-reactive carcinogens, Ashby and Paton[41] determined that, with the exceptions of chemicals or exposures which were not appropriate to classify in this manner (*e.g.* asbestos), and agents related to hormone-mediated carcinogenesis, the remainder of the acknowledged human carcinogen exposures were limited to the DNA-reactive rodent carcinogens. Their analysis of the non-DNA-reactive rodent carcinogens suggested that the observed carcinogenesis was more related to the experimental species than to the test chemical.

The foregoing concerns with mechanisms of carcinogenesis have two major consequences for cancer risk assessment.[1] First, substances which test positive in the rodent bioassays may operate by a mechanism which does not apply to humans because of anatomical, physiological or biochemical differences between species. Second, the substance may test positive in the rodent bioassay under the high dose conditions of the experiment by some means of either direct or indirect tumour promotion such as cytotoxicity or alterations in hormonal or immune responses. However, these tumour promotion mechanisms may demonstrate a threshold because their effects at low dose can be compensated by repair or homeostatic mechanisms. This compensation can make the application of the no-threshold linear model seriously overestimate the cancer risk at low environmental doses.

For the first case, several tumour mechanisms have been identified which appear to be unique to rodent species.[42,43] These include:

- Kidney tumour formation in male rats caused by (+)-limonene, which binds with endogenous protein
- Tumour formation by butylated hydroxyanisole caused by accumulation in the rodent forestomach
- Bladder tumour formation in aged male rats caused by high doses of saccharin beginning in the perinatal period, ultimately causing an irritant precipitate
- Various rodent liver tumours attributed to peroxisome proliferation

In the second case, a number of epigenetic (non-genetic) mechanisms, including

[40] S. M. Cohen and L. B. Ellwein, *Science*, 1990, **249**, 1007.

[41] J. Ashby and D. Paton, *Mutat. Res.*, 1993, **286**, 3.

[42] International Expert Panel on Carcinogen Risk Assessment, American Health Foundation, *Pharmacol. Therapeut.*, 1996, **71**, (1/2), 1.

[43] J. Ashby, A. Brady, C. R. Elcombe, B. M. Elliot, J. Ishmael, J. Odum, J. D. Tugwood, S. Kettle and I. E. H. Purchase, *Hum. Exp. Toxicol.*, 1994, **13** (suppl. 2), S1.

disproportionate liver tumour response, have been identified for various agents which are carcinogenic in rodent bioassays and which may or may not be relevant to humans.[44,45] Although not all of the epigenetic mechanisms can be ruled out as having human relevance, epigenetic mechanisms generally suggest the existence of a threshold for their effects. Although cell proliferation has been observed to play a role in epigenetic mechanisms of carcinogenesis, the ability of an agent to enhance cell division by itself chemically has not been a reliable predictor of carcinogenesis in long-term bioassays.[46]

Emerging knowledge of the roles of additional processes in carcinogenesis has resulted in the development of more complex models which attempt to address the various processes believed to be operating.[40,47] These models are useful for developing our understanding of the mechanisms involved in different experimental findings, but they are still limited in their ability to extrapolate to low doses because some of the major input parameters cannot be measured independently and substantial uncertainty remains about the details of various mechanisms involved.[29] Consequently, despite the conceptual utility of these more complex models, they do not allow confident estimation of expected cancer risk at extrapolated low dose.

Although there has been a virtual explosion of knowledge about molecular and cellular level mechanisms in cancer, the quantitative risk assessment model has remained remarkably crude. This situation reflects the enormous difficulty of the challenge originally presented to quantitative cancer risk assessment and explains the need to restrict applications of quantitative estimates to bounding the cancer risk potential rather than attempting to determine expected cancer incidence. Yet this bounding role should still allow for evidence to emerge which can eliminate substances from substantive suspicion of being human carcinogens, based upon evidence of mechanisms which will not pose a risk for humans. This possibility was foreshadowed in the scientific foundations for the 1986 EPA guidelines:[28] 'As science advances, and more sophisticated studies are done, it may be possible to demonstrate the existence of unique susceptibilities to chemical carcinogenesis in rodents that have no relevance for humans'.

In summary, our emerging knowledge of cancer mechanisms is showing an exceedingly complex array of interactions rather than the very simplistic concept inherent in the linear, no-threshold regulatory model of carcinogenesis which was largely predicated on a single carcinogen causing a mutation which ultimately develops into a tumour according to a random stochastic process.

Maximum Tolerated Dose and Experimental Protocols

Originally, rodent cancer bioassays were developed as a qualitative or at most semi-quantitative tool for assessing the potential of a substance to cause cancer.

[44] J. I. Goodman, J. M. Ward, J. A. Popp, J. E. Klaunig and T. R. Fox, *Fundam. Appl. Toxicol.*, 1991, **17**, 651.

[45] R. H. Alison, C. C. Capen and D. E. Prentice, *Toxicol. Pathol.*, 1994, **22**, 179.

[46] R. L. Melnick, J. Huff, J. C. Bennett, R. R. Maronpot, G. Lucier and C. J. Portier, *Mol. Carcinogen.*, 1993, **7**, 135.

[47] S. Moolgavkar and A. G. Knudson, *J. Natl. Cancer Inst.*, 1981, **66**, 1037.

Figure 1 Association between the upper bounds on low dose slope estimates and maximum dose used in rodent carcinogen bioassays. (After ref. 48)

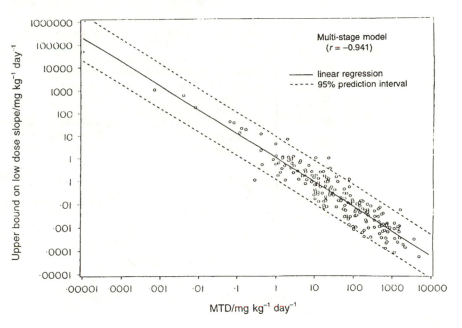

With the demands for quantitative risk assessment came demands for more animal data on more substances. This led to the development of exhaustive protocols for these long-term studies (generally two years) whose cost rose into the millions of dollars. With the level of investment involved and the practical limitations on number of animals at each dose level, choosing the MTD as the top dose level was a pragmatic decision to maximize the chance of the experiment to detect a carcinogenic response. This position was easier to rationalize if results were used in strictly a qualitative manner, although our increased understanding of epigenetic mechanisms have raised important qualitative concerns about tumour responses at MTD.

Considering these issues, the possibility of the extrapolated linear dose–response slope being influenced by the study protocol was a concern. In most cases, the bioassay dose regimes are the MTD, MTD/2 and the MTD/4. Accordingly, the MTD is usually the maximum dose tested. Krewski *et al.*[48] considered explicitly the relationship between the maximum dose used and the upper bound estimate on the linear slope for the multistage model (q_1^*) for 191 different carcinogens and found there was a remarkable negative correlation $r = -0.941$ (Figure 1). This was explained by noting that the nature of the study design only allowed a 22-fold range in extrapolated slope estimates once the maximum dose was determined or only an 11-fold range for slope if the lowest dose (MTD/4) was not tested. In other words, determination of the MTD, which is strongly influenced by the acute toxicity of a substance, largely determines the value of the extrapolated carcinogenic slope. This finding is understandable because the lower end of the linear slope line is constrained to pass through zero, so the more

[48] D. Krewski, D. W. Gaylor, A. Soms and M. Szyskowicz, *Risk Anal.*, 1993, **13**, 463.

Figure 2 Schematic of rodent carcinogen bioassay dose–response curve illustrating constraints on upper bound slope estimates, q_1^*

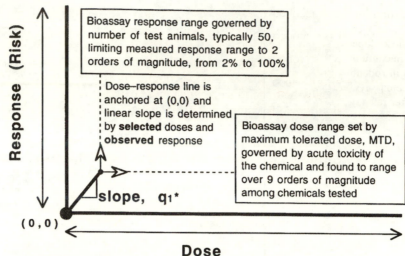

toxic a substance (MTD closer to zero), the steeper the resulting slope must be (Figure 2).

The foregoing observations suggest that the major quantitative parameter derived from chronic cancer bioassays is largely dictated by the maximum dose used in the bioassay and the protocol itself. This revelation raises serious questions about the degree of insight into low dose cancer risk which is provided by these quantitative methods. If nothing else, it surely reinforces the earlier message that these methods are incapable of generating meaningful estimates of expected cancer risk and can only be viewed as providing some upper bound limits on cancer risk.

Another emerging concern with the protocol has been the finding that feeding regimes in chronic cancer bioassays can be very important to observed tumour incidence.[49] The standard bioassay protocols has allowed rodents to feed at will and, over the years, the average body mass of rodents has been increasing. This change has been found to have a strong impact on tumour incidence, with body mass increases as small 15% leading to a doubling of liver tumours in male mice from a rate of 10% to 20%.[50] Consequently, consideration must be given to introducing dietary restriction or dietary control to avoid having this experimental factor further undermine the validity of the test results.

Compounded Conservatism and Confronting Uncertainty

The original intention of providing plausible upper bound estimates of cancer risk was used to justify the selection of parameters which were intended to

[49] International Life Sciences Institute, *Dietary Restriction: Implications for the Design and Interpretation of Toxicity and Carcinogenicity Studies*, ed. R. W. Hart, D. A. Neumann and R. T. Robertson, ILSI Press, Washington, DC, 1995.

[50] A. Turturro, P. Duffy and R. W. Hart, *Dietary Restriction: Implications for the Design and Interpretation of Toxicity and Carcinogenicity Studies*, ed. R. W. Hart, D. A. Neumann and R. T. Robertson, ILSI Press, Washington, DC, 1995, pp. 77–97.

overestimate the cancer risk. They have attracted considerable debate about how plausible they were. Individual values used in the risk calculations were often selected to be the 95% upper (or lower, depending on the effect of a parameter on the risk calculation) bound estimate. The intent was to estimate something approaching the 95% upper bound on the risk estimate but the effect would usually be to produce a much more extreme upper bound estimate because the simultaneous combination of several rare value inputs yields an even more exceptionally rare value output.

This phenomenon of compounded conservatism has been illustrated by employing the technique of Monte Carlo simulation, which allows the use of distributions of values for input parameters to calculate an output distribution for the risk calculation.[51] One example calculation showed that using upper 95% point estimate values for all input parameters yielded a single-value output risk calculation which was almost 400-fold higher than the upper 95% value on the full output risk distribution.

Policies vs. Facts

The enormous gaps in our knowledge about the complexities of carcinogenesis and procedures for quantitatively estimating cancer risk from animal experiments has to be bridged by making many critical assumptions. In most cases, available knowledge is not adequate to select among those assumptions so that the choices have had to be guided by policy. There are too many of these policy choices involved to address them all within the confines of this article. However, the important insight is to distinguish these policy choices, made under an umbrella of conservatism, from scientific facts. While those closely involved in making risk assessment decisions may have appreciated the important distinction, the wider communication of this distinction among users and the public has been inadequate.

Perhaps the most compelling example of this problem surrounds the issue of assuming no threshold for a carcinogen. This assumption, predicated on the remote, but conceivably non-zero, possibility of one molecule of a carcinogen initiating a tumour, has pervaded quantitative risk assessment practice to the extent that some may be forgiven for believing it to be a scientific fact. For example, an astounding 58% of Danish medical students and 66% of Danish risk assessment students either agreed or strongly agreed with the statement:[52] 'There is no safe level of exposure to a cancer-causing agent'. This premise was enshrined in some of the early policy statements on cancer risk assessment, as noted in Section 2. Yet, even if one accepts the cautious upper bound policies, Hrudey and Krewski[53] demonstrated that the upper bound cancer risk estimate using the steepest slope factor estimated for any carcinogen (2,3,7,8-TCDD) applied to a dose of one molecule a day for a lifetime yielded an infinitesimal risk estimate. If applied to the entire population of the planet, less than 0.00001 case of cancer would be predicted. While accepting that safety is a value-laden, individual concept, it seems difficult to label this level of risk, for the smallest conceivable

[51] T. E. McKone and K. T. Bogen, *Environ. Sci. Technol.*, 1991, **25**, 1674.
[52] P. Grandjean and J. B. Nielsen, *Ugeskr. Laeger*, 1996, **158**, 5291.
[53] S. E. Hrudey and D. Krewski, *Environ. Sci. Technol*, 1995, **29**, 370A.

lifetime chronic dose of a carcinogen, as being unsafe. Of course, it is much more challenging to decide how much higher exposure could go before being considered unsafe, but it is important to recognize that a safe level, by any practical notion of safety, is conceivable. Because other carcinogens are much less potent, if a safe dose for TCDD is conceivable, it follows that there can be a safe dose for any other.

Policy in risk assessment has created many other difficulties. Because different agencies in government developed their own policies on how to resolve various scientific choices, different risk estimates were often generated from the same set of animal evidence. While it might seem to be simply an issue where some coordination is required, in the US, each regulatory agency is subject to legal challenges on its decision, so maintaining consistency with previous decisions becomes an important issue or precedent which discourages agencies from changing policies to achieve harmony.[10]

Environmental Chemicals and Cancer

The expectation that regulating chemical carcinogens in the environment could achieve a major reduction in the burden of cancer in society was a major motivation for developing cancer risk assessment. However, while these measures were being proposed with increasing frequency, other views were emerging to refute the contention that chemical carcinogens in the environment were a major factor in human cancer. Dr John Higginson, the founding director of IARC and a leading proponent of the preventable nature of cancer, spoke out to clarify what he saw as the misinterpretation of his extensively cited theories about the role of environment in carcinogenesis.[54] He noted that much of the confusion surrounded his meaning of environment, which included every aspect of living. In contrast, many had interpreted environment to mean only chemicals, often only man-made chemicals.

The most influential analysis on this issue was published by Doll and Peto[55] in 1981. They provided an exhaustive review of epidemiologic evidence on the causes of cancer and estimated that pollution might be responsible for < 1% to 5% of all cancer deaths in the US, with their best estimate at 2%. They attributed the major causes to tobacco at 30% (25–40%) and diet at 35% (10–70%). A later analysis attempted to compare these with estimates using the EPA methodology.[56] These were summed for several pollution categories to yield a total estimate of between 1% and 3% of all cancer deaths. Even though these are upper bound estimates, they are consistent with Doll and Peto in suggesting that environmental pollution is not a major factor in human cancer.

Opportunities for misunderstanding on the cancer issue also arise from the manner in which cancer statistics are viewed. Because the population is increasing, the total number of cancer cases will increase, so comparisons over

[54] J. Higginson, *Science*, 1979, **205**, 1363.
[55] R. Doll and R. Peto, *J. Natl. Cancer Inst.*, 1981, **66**, 1192.
[56] M. Gough, *Environ. Sci. Technol.*, 1989, **23**, 925.

time must be based upon rates per unit of population.[57] Even more important, cancer incidence and mortality increases substantially with age, so meaningful comparisons over time requires the use of age-standardized rates which adjust the crude rates to a common age distribution. When viewed on an age-standardized basis, the only major cancer type showing increasing mortality over the past 20 years in Canada is prostate cancer for males and lung cancer for females.[3] Lung cancer had been increasing in men until the mid-1980s, but has begun to decline because of the reduction in smoking among men. The continuing increase of lung cancer in women unfortunately reflects their later adoption of this habit. Even with the extensive information on the incidence and mortality from cancer, making connections with causative agents through indirect evidence and inference from epidemiology and through the predictive inference of toxicological risk assessment is a vastly more uncertain source of knowledge than the basic cancer statistics.[57]

Another challenge to the importance of the synthetic chemical carcinogen theory was mounted by Bruce Ames and colleagues,[58,59] who noted that about half of the natural pesticides which had been tested for carcinogenicity were positive in animal experiments. Furthermore these natural pesticides, produced by plants themselves, constitute 99.99% of the weight of pesticides consumed in the American diet. Looking back on the expectations for carcinogen regulation *versus* subsequent experience with cancer, the founder of the US EPA risk assessment approach acknowledged that the expected reductions in cancer rates have not materialized.[10]

5 Advances and New Approaches for Cancer Risk Assessment

There has been a virtual explosion of information over the past decade emerging from the massive research effort directed at causes and treatment of cancers. These findings have varying degrees of relevance to quantitative cancer risk assessment and it is not possible to provide an exhaustive review of all of these developments within the confines of this article. Consequently, the following is a sampling of some major initiatives which have a direct bearing on the future practice of quantitative risk assessment.

Hazard Identification and Cancer Mechanisms

The main advances relevant to the quantitative stage of cancer risk assessment have been the growing understanding about mechanisms of cancer development, particularly as they relate to tumour development in experimental animals and their resulting relevance to human cancer risk. Important advances involve recognizing and documenting epigenetic mechanisms of tumour development, as discussed in Section 4. Some mechanisms are clearly unique to particular species of rodent and are, apparently, of no relevance to human cancer risk. Other

---

[57] S. Thomas and S. E. Hrudey, *Risk of Death in Canada—What We Know and How We Know It*, University of Alberta Press, Edmonton, 1997.

[58] B. N. Ames, M. Profet and L. S. Gold, *Proc. Natl. Acad. Sci. USA*, 1990, **87**, 7777.

[59] B. N. Ames and L. S. Gold, *Environ. Health Persp.*, 1997, **105**, (suppl. 4), 865.

mechanisms may occur in humans, but only at the high dose conditions used in rodent bioassays. Because many, if not most, of these mechanisms likely involve a threshold, below which cancer risk can be dismissed, application of no-threshold extrapolation models for such carcinogens is misleading and inappropriate.

The importance of identifying and distinguishing epigenetic mechanisms of carcinogenesis has been stressed by multiparty expert reviews of past practice.[27,29,42] Table 3 provides a summary of appropriate evidence of identifying epigenetic mechanisms with sufficient certainty to abandon the conservative default assumption of a carcinogen being genotoxic.[42]

Research has also illuminated a clearer role for genetic factors in cancer initiation and development. The premise that certain genes, so-called oncogenes, had to be mutated to initiate tumours was demonstrated by finding that introduction of the *ras* oncogene into primary cells can initiate normal cells on the pathway to tumour formation.[60] The correspondence of these specific mutations with the tumorigenic action of known genotoxic carcinogens has been demonstrated.[61] Yet the genetic relationships involved in tumour development are complex, as illustrated by the important role of the p53 tumour suppresser gene, whose inactivation by mutation has been noted in 60% of human tumours.[62] So, genetic factors involve not only the initiation of the process, but also the progression of tumours in initiated cells because of genetic impairment of normal cellular process control. The spectrum of p53 mutations differs among different cancer types, suggesting a mix of internal and external factors in the genetic factors involved in human carcinogenesis.[63] The emerging genetic

Table 3 Evidence for identification of an epigenetic mechanism	*Lack of genetic mechanism*	No relevant DNA adducts identified
		Negative results in genotoxicity assays
	Evidence of epigenetic mechanism	Causes cell damage in target organ
		Enhanced cell proliferation, especially of initiated cells
		Causes tumour promotion following initiating carcinogen
		Related biochemical or physiological alterations
	Dose- and time-related effects	Requires a prolonged high dose
		Effects are initially reversible
		Thresholds for epigenetic effects
		Doses producing epigenetic effects $\leq$ doses producing tumours

[Reprinted with permission from *Pharmacol. Ther.*, 1996, **71** (1/2), 1, Elsevier Science Inc., 655 Avenue of the Americas, New York NY 10010-5107.]

[60] R. A. Weinberg, *Cancer Res.*, 1989, **49**, 3713.
[61] S. Sukumar, *Cancer Cells*, 1990, **2**, 199.
[62] A. J. Levine, M. E. Perry, A. Chang, A. Silver, D. Dittmer, M. Wu and D. Welsh, *Br. J. Cancer*, 1994, **69**, 409.
[63] M. Hollstein, D. Sidransky, B. Vogelstein and C.C. Harris, *Science*, 1991, **253**, 49.

complexities of carcinogenesis make overly simplistic quantitative models of tumour initiation even more indefensible for making expected cancer risk estimates.

The majority of information available to hazard identification has been and continues to be from experimental animals, despite the widely accepted premise that the most credible evidence of human carcinogenicity can be obtained from epidemiologic investigations. This apparent paradox arises for several reasons. Ethical obligations preclude experimental human exposures. Analytical epidemiologic studies necessary to contribute to causal inference are:

- Very expensive
- Delayed by the need for decades of follow-up after exposure
- Subject to difficulties in documenting exposures, controlling bias and confounding
- Only sensitive to comparative risk levels many orders of magnitude higher than regulatory levels, so that risks must still be extrapolated
- Unable to provide predictions of risk from a hazard until humans have been substantially exposed.

However, many large-scale epidemiologic studies have been undertaken in the last decade and there is a need to make maximal use of the best available evidence from these studies. To this end, an expert panel enunciated a set of principles for evaluating epidemiologic data for risk assessment.[64] These addressed the merits of a study for evidence of a causal relationship by means of a checklist of questions testing each principle.

The experience of two decades of cancer risk assessment practice and an expert review led to a long list of recommendations to the US EPA to revise its 1986 guidelines.[27] A new draft, released for discussion, calls for a number of changes.[65] One major issue was a call for a clearer elaboration of default assumptions, the rationale for them and criteria for departing from them.[27]

Another major change proposes to increase the role of evidence about an agent's properties, its structure–activity relationship to known carcinogens and its behaviour in biochemical studies of carcinogenesis at the cellular level for contributing to the weight of evidence in classifying a carcinogen. Classification should also consider the nature of a carcinogen's potential, such as distinguishing an agent which may be carcinogenic by inhalation but not by ingestion (*e.g.* asbestos). Even more striking is a recommendation to eliminate use of the alphanumeric ranking scheme (Table 2). Rankings will be replaced with a narrative based upon elaboration of the descriptors 'known/likely', 'cannot be determined' and 'not likely'. This major departure from prior practice was justified: to permit more explanation of data strength and limitations, to avoid an even more complex coding system necessary to accommodate the expanding range of qualifiers appropriate to hazard identification, and to provide a more flexible system to accommodate future advances in knowledge arising from the course of cancer research.

[64] Federal Focus, Inc. *Principles for Evaluating Epidemiologic Data in Regulatory Risk Assessment*, Federal Focus Inc., Washington, DC, 1996.

[65] US EPA, *Proposed Guidelines for Carcinogen Risk Assessment*, Office of Research and Development, US Environmental Protection Agency, EPA/600/P-92/003C, Washington, DC, 1996.

Dose–Response Assessment

The default position for time course scaling of carcinogen dose between experimental animal and human exposures has been to apportion a single high dose in humans as if it had been delivered incrementally on a daily basis. While this assumption has long been recognized as tenuous, alternative approaches have generally not had sufficient data to demonstrate their utility. A Dutch analysis of relevant animal studies involving single or short-term exposures was performed to determine possible ranges of values for a dose rate correction factor (DRCF), defined as the factor by which the tumour incidence at low dose rates is multiplied to derive tumour incidence at high dose rates.[66] A range of DRCF from 0 to 8.3 was derived with the qualifier that high or low values would be dependent upon the operative mechanisms. In particular, peak exposure to a carcinogen affecting an early stage of carcinogenesis would be expected to have its greatest effect early in life, while peak exposure to a carcinogen affecting a late stage of carcinogenesis would be expected to have its greatest effect late in life.

Another perspective on dose rates came in an analysis of rodent bioassay and human epidemiologic evidence to consider alternative dose rate measures for 2,3,7,8-TCDD. This demonstrated, using three measures of cumulative lifetime dose for this slowly eliminated contaminant, that humans may be from 9 to 90 times less responsive to TCDD than rats.[67]

The past decade has seen enormous advances in the practice of physiologically based pharmacokinetic (PB-PK) modelling in toxicology. This has advanced the field of pharmacokinetics, which had relied primarily on empirical models which offered little insight into mechanisms of uptake, distribution, metabolism and excretion of toxic agents, and did not facilitate the translation of observations from experimental animals to humans.[68] So, rather than scaling doses from animals to humans based on size, physiologically based models study the pharmacokinetics in the experimental species and scale the dose to humans using relevant determinants of the pharmacokinetics (blood flow rates, tissue volumes and rates of metabolism). This approach has been applied to methylene chloride to show that strictly empirical scaling can substantially overestimate (21-fold) the dose of active agent delivered to the human tissue site of tumour formation.[69] The enormous potential of PB-PK modelling for refining biologically relevant dose estimates will have continuing impact on quantitative cancer risk assessment as PB-PK models for more carcinogenic agents are developed and validated. Furthermore, future improvements can be expected as knowledge develops about differences across species in tissue sensitivity for carcinogenic response to the delivered dose of active agent.[70]

The mounting doubts about the meaning of numbers generated by low dose

[66] H. Verhagen, V. J. Feron and P. W. van Vliet, *Risk Assessment of Peak Exposure to Genotoxic Carcinogens*, Health Council of the Netherlands, The Hague, 1994, Pub. no. A94/04.

[67] L. L. Aylward, S. M. Hays, N. J. Karch and D. J. Paustenbach, *Environ. Sci. Technol.*, 1996, **30**, 3534.

[68] K. Krishnan and M. E. Andersen, *New Trends in Pharmacokinetics*, ed. A. Rescigno and A. K. Thakur, Plenum Press, New York, 1991, pp. 203–226.

[69] K. Krishnan and M. E. Andersen, in *New Trends in Pharmacokinetics*, ed. A. Rescigno and A. K. Thakur, Plenum Press, New York, 1991, pp. 335–354.

[70] M. E. Anderson, H. Clewell and K. Krishnan, *Risk. Anal.*, 1995, **15**, 533.

risk extrapolation have caused reconsideration of the merits of this approach for deriving regulatory limits. A British proposal suggests that it may be possible to define either a No Expected Human Effect Level from epidemiologic studies or a No Expected Animal Effect Level from animal bioassays.[71] These would represent thoughtful analyses of available data on genotoxic carcinogens to identify exposure levels which appear to be too low to demonstrate an effect by any practical and reliable measurement. Then, without debating the existence or absence of a threshold, adjustment factors would be applied to define an even lower exposure level which would not be expected, with a suitable margin of confidence, to produce any observable harm in humans.

The assessment of carcinogenic risk of priority substances under the Canadian Environmental Protection Act (CEPA) followed a slightly different approach.[72] While acknowledging that genotoxic carcinogens may have no threshold, the method defines a Tumorigenic dose 05 (TD_{05}), the dose associated with a 5% increase in incidence or mortality due to tumours, after scaling for interspecies variation. The TD_{05} is derived from a dose–response curve fit, extended to the 5% tumour response level. This would be slightly beyond the lowest level ($\sim 10\%$) normally observable in a typical rodent bioassay. Then a margin of exposure approach would be applied to reduce the TD_{05} to a tolerable level by dividing it by adjustment factors, currently under development. Health Canada notes that for the carcinogens evaluated under CEPA, adjustment factors in the range of 5000 to 50 000 would provide criteria levels in the same range as those carrying extrapolated lifetime cancer risk estimates of 10^{-5} to 10^{-6}.

The proposed new US EPA guidelines would replace the LMS model as the default, noting that this model 'gave an appearance of specific knowledge and sophistication unwarranted for a default'.[65] This comment was probably directed at the misunderstanding among some users that the LMS model was a mechanistic rather than an empirical model because its name mentions the commonly accepted premise of cancer being a multistage process. The new proposal is similar to the Canadian proposal in the sense of determining a dose level corresponding to a specified tumour response by fitting the dose–response data. However, it then specifies choosing the intersection of the upper 95% interval on the dose–response curve fit with the 10% response level to define a lower bound dose, designated the LED_{10}. This value is similar to that for non-carcinogens using the benchmark dose approach.[73] At this stage, the new proposal would, as a default, then extend a straight line to zero dose, zero response for extrapolating the low dose cancer risk. Only if biological evidence can justify a threshold mechanism of carcinogenesis, say a tumour promotion mechanism, would a margin of exposure approach (division of the LED_{10} by adjustment factors as in the Canadian method) be used to determine a tolerable

[71] R. L. Maynard, K. M. Cameron, R. Fielder, A. McDonald and A. Wadge, *Hum. Exp. Toxicol.*, 1995, **14**, 175.

[72] M. E. Meek and G. Long, *Health-Based Tolerable Daily Intakes/Concentrations and Tumorigenic Doses/Concentrations for Priority Substances*, Environmental Health Directorate, 96-EHD-194, Health Canada, Ottawa, 1996.

[73] K. Crump, B. Allen and E. Faustman, *The Use of the Benchmark Dose Approach in Health Risk Assessment*, Risk Assessment Forum, US Environmental Protection Agency, Washington, DC, 1995.

dose. Otherwise, the proposed US EPA guidelines will retain the means to calculate an estimate of low dose risk using the extrapolated straight line slope.

Exposure Assessment

Substantial improvements in exposure assessment have arisen by means of advances in modelling approaches, particularly the availability of Monte Carlo techniques for incorporating distributions of exposure-related parameters into the risk assessment determinations.[74] These approaches have allowed progress beyond the default, point-estimate approaches inherent in earlier risk assessment protocols. They allow development of an appreciation of the shape of an outcome distribution and the input factors which primarily determine its shape.

Because of the growing use of these methods, the US EPA recently released a policy statement on the use of probabilistic analysis in risk assessment.[75] This policy accepted these methods for exposure assessment in regulatory risk assessments, provided that eight conditions were satisfied. These conditions mainly related to providing sufficient documentation to identify the sources and relevance of input data and methods to allow reproduction of the analysis.

A major advance in comprehensive screening models for risk assessment has been achieved with the development of the CalTox model which links a multi-media fugacity model for tracking the fate and behaviour of contaminants with an exposure model and an extensive database of contaminant properties.[76,77] This model, designed specifically for contaminated site evaluations, is built on a standard spreadsheet and can be accessed on the worldwide web at: http://www.cwo.com./~herd1/downset.htm

Risk Characterization, Management and Communication

An important benefit of the increasing use of probabilistic methods in risk assessment has been recognition of the distinction between variability and uncertainty.[78] Variability arises when there are truly different values for a given parameter within a population or a sample. For example, there are true variations in concentrations of contaminants, breathing rates, body mass and susceptibility to contaminants. These will result in variations in the level of risk to any individual. The range of variation for any parameter can be represented by a probability distribution. Consequently, Monte Carlo analyses have been useful for bringing variability into exposure assessments.

Confusion has arisen from a failure to distinguish uncertainty from variability, perhaps because uncertainty can also be represented by a probability distribution.

[74] B. L. Finley and D. J. Paustenbach, *Risk Anal.*, 1994, **14**, 53.

[75] US EPA, *Policy for Use of Probabilistic Analysis in Risk Assessment*, Office of Research and Development, US Environmental Protection Agency, EPA/630/R-97/001; internet at http://www.epa.gov/ncea/mcpolicy.htm, 1997.

[76] The Office of Scientific Affairs, *Caltox—A Multimedia Total Exposure Model for Hazardous Waste Sites*, Department of Toxic Substances Control, California Environmental Protection Agency, Sacramento, CA, 1993.

[77] T. E. McKone and J. I. Daniels, *Regul. Toxicol. Pharmacol.*, 1991, **13**, 36.

[78] D. Hattis and D. E. Burmaster, *Risk Anal.*, 1994, **14**, 713.

True uncertainty is about how confident we are about our knowledge of any particular value to be used in a risk calculation. Uncertainty can arise from many sources, including sampling error, inaccurate measurement methods or inaccurate concepts or models. So uncertainty relates to how sure we can be about the risk for any particular individual, whereas variability determines how risk may differ from one individual to another within the population.

Because Monte Carlo methods simply process input probability distributions, applications of this method to risk assessment have often mixed variability and uncertainty analyses together with no attempt to discriminate the concepts. This is a concern, because they represent different problems and different solutions. Variability in risk cannot be reduced by further study; it can only be better characterized to understand the differences in risk exposures. Uncertainty in risk can, at least in theory, be reduced if research can improve our understanding of the item being measured.

The issue of variability and uncertainty in risk characterization has been eloquently addressed by Hattis and Barlow,[79] who evaluated evidence on interindividual variability among a number of relevant processes for genotoxic carcinogens, including metabolic activation, detoxification and DNA repair. Based on their analysis, they demonstrated how risk managers should consider risk in three dimensions: the level of risk, the proportion of a population at that risk level because of interindividual variability and the degree of confidence in the risk estimate because of true uncertainty.

The proposed new EPA guidelines stress incorporating greater discussion of the extent and weight of evidence, including strengths and weaknesses for the assessment of a given agent.[65] The required discussion should also stress the interpretation and rationale used, provide alternative conclusions and address the major uncertainties inherent in the analysis.

The process of quantitative cancer risk assessment has been consistently driven to answer the original two questions about whether a substance is a human carcinogen, and if it is, how large a risk does it pose? Over the last two decades, the cancer risk assessment process has been applied to an increasing range of problems, often with the tacit assumption that providing answers to these two questions will resolve the problem at hand. A recent expert review of the risk characterization process has questioned the validity of this assumption.[80]

If risk characterization is to bridge what can be known about a problem to decide what should be done about it, then risk characterization must be able to inform the decision-making process. So, although it is essential for risk assessment to deal with all of the scientific issues, risk assessment aimed at informing decision-making must involve the major parties to a risk issue in formulating the problem to be addressed. Furthermore, an effective process cannot be a once-through endeavour which delivers a final and absolute pronouncement characterizing the risk. That model presumes that a once-through risk characterization, even if it acknowledged the enormous uncertainties involved, would be able to dictate a 'correct' course of action. Rather, risk

[79] D. Hattis and K. Barlow, *Hum. Ecol. Risk Assess.*, 1996, **2**, 194.
[80] National Research Council, *Understanding Risk: Informing Decisions in a Democratic Society.* ed. P. C. Stern and H. V. Fineberg, National Academy Press, Washington, DC, 1996.

assessment which will effectively inform risk management decisions must be an iterative process. As more understanding of a problem is gained, efforts can be directed at refining risk characterization or elaborating uncertainty in those areas most important for resolving the problem at hand.

Summary

Quantitative cancer risk assessment was born out of the controversy surrounding an attempt by US EPA lawyers to enshrine a set of 'principles' to capture the consensus of what was known about cancer. The two decades of practice have caused many more controversies but, recently, a new set of 'principles' on cancer have been published.[42] These statements, provided in Table 4 summarize much of the discussion in this review and express the consensus among many current international experts on cancer risk assessment.

6 Conclusions

1. Quantitative cancer risk assessment is an important aid to environmental decision-making which was born amid intense controversy and has been struggling with misunderstanding ever since. Many of the controversies can be resolved by recognizing the inherent limitations to the methods which are constrained by what we can and do know about cancer. This requires keeping expectations for these methods within the bounds of what they can possibly deliver.

2. Despite several explicit cautions built into various practice guidelines, the risk estimates produced have often been misinterpreted as 'expected' cancer risks rather than cautious overestimates seeking to bound the plausible cancer risk. The resulting misunderstandings have seriously reduced the credibility of the process within the science community as well as with the public.

3. There has been enormous progress in our understanding of cancer mechanisms and this knowledge can be used to be much more discriminating in applying the results of animal cancer experiments to estimating human cancer risk.

4. For those matters which cannot be resolved by scientific knowledge, policy choices will still need to guide risk assessment practice, but such policies need to be clearly distinguished from facts.

5. The early expectations of major achievements in reducing human cancers by strictly regulating carcinogens in the environment were based on a misconception about the role of environment in cancer. Current understanding would estimate the role of environmental pollution by chemical carcinogens as causing a very small percentage of total human cancers. However, because cancer is so prevalent in society, a substantial number of environmentally related cancer cases may still be preventable. Localized higher level exposures to chemical carcinogens should be reduced and quantitative cancer risk assessment has an important role in identifying priorities for meaningful prevention of exposures.

6. Some of the controversy created by misunderstanding of risk assessment predictions might be avoided if more attention was invested in placing the bounding risk predictions into a broader context of what epidemiologic studies can tell us about cancer risks in society.

Table 4 Principles of carcinogenesis and risk assessment for chemicals

1. A carcinogen is operationally defined as an agent that is causally related to neoplastic transformation, leading to malignant neoplasms, or benign neoplasms that may progress to malignancy
2. Carcinogenesis is a complex, multi-event process. Chemicals may be involved in DNA damage and mutation, modification of biochemical and cellular processes, alteration of proliferation in normal cells, enhancement of proliferation in mutated cells, progression of cells to a malignant state or growth of malignant tissue
3. Genetic alteration in cellular DNA, which may be inherited or due to endogenous or exogenous factors, is necessary for neoplastic conversion
4. The process of carcinogenesis is modulated by a variety of intrinsic factors, including genetic background, age, cell proliferation, enzyme balance, hormonal status and immunological competence, as well as extrinsic factors such as diet
5. Many biological consequences of DNA damage, *e.g.* mutation, are irreversible and result in cells exhibiting progressive autonomy, whereas some consequences of epigenetic changes, *e.g.* hyperplasia, remain under homeostatic control for a period of time and evidence reversibility upon withdrawal of the chemical
6. Carcinogens act by various mechanisms and demonstrate different potencies. Some act primarily through DNA reactivity; others act exclusively by epigenetic mechanisms. DNA-reactive carcinogens can also exert epigenetic effects
7. Carcinogenic effects observed in experimental animals are considered relevant to assessing carcinogenic effects in humans, except where the mechanism by which tumours are produced in animals do not apply to humans
8. Epigenetic carcinogens act through a variety of mechanisms, which may demonstrate thresholds. Such agents are expected to pose no carcinogenic risks at sub-threshold exposures
9. DNA-reactive carcinogens produce heritable alterations in genetic material, which may occur even at low exposure levels. At exposures below which saturation of metabolic processes occurs, excess cancer risk may be proportional to exposure
10. Carcinogenic risk assessment should be done on a case-by-case basis, according to the biological mechanisms underlying carcinogenesis.

[Reprinted with permission from *Pharmacol. Ther.*, 1996, **71** (1/2), 1, Elsevier Science Inc., 655 Avenue of the Americas, New York NY 10010-5107.]

7. Perpetuating a view that there can be no safe level of exposure to any carcinogen makes societal notions of safety unachievable at best and meaningless at worst.

8. Conservatism in estimating cancer risk can be justified as a cautious public policy, but there is a need to know how cautious adopted policies are and how they will deal with variability in risk exposure and uncertainty in risk magnitude.

9. Quantitative cancer risk assessment should be committed to informing decision-making, which demands that all parties to a decision should be involved in formulating an understanding of the problem to be assessed. Having risk

assessment specialists develop a routine, standardized technical prediction of a cancer risk in isolation and then attempt to explain its meaning to decision makers and affected parties is only likely to generate scepticism and distrust.

10. As in the beginning, the key to major advancement in quantitative risk assessment must be a better understanding of causes and mechanisms of cancer. However, it is important that as our knowledge continues to improve, better means are found to incorporate that knowledge into risk assessment practice.

7 Acknowledgements

The inspiration, guidance and healthy scepticism that Professor Roger Perry instilled in his first research student at Imperial College is gratefully acknowledged. The substantial assistance of Nola Low in tracking and acquiring literature sources for this review and the careful editorial assistance of Elizabeth Hrudey is appreciated. This work was supported by funding for the Eco-Research Chair in Environmental Risk Management provided by the Tri-Council Secretariat representing the Medical Research, the Natural Sciences and Engineering Research and the Social Sciences and Humanities Research Councils of Canada; the Alberta Heritage Foundation for Medical Research; Alberta Environmental Protection; Alberta Health; the City of Edmonton; Syncrude Canada Ltd.; and the Alberta Energy and Utilities Board.

Notation

CAG	=	Carcinogen Assessment Group
CEPA	=	Canadian Environmental Protection Act
DRCF	=	dose rate correction factor
FDA	=	Food and Drug Administration, USA
IARC	=	International Agency for Research on Cancer
LED10	=	lower bound of 10% effective dose
LMS	=	linearized multistage model
MEI	=	maximally exposed individual
MLE	=	maximum likelihood estimate
MTD	=	maximum tolerated dose
NCI	=	National Cancer Institute, USA
PB-PK	=	physiologically based – pharmacokinetic
RME	=	reasonable maximum exposure
RSD	=	risk specific dose
TCDD	=	tetrachlorodibenzo[p]dioxins
TD05	=	dose for 5% tumourigenic response
US EPA	=	US Environmental Protection Agency
VSD	=	virtually safe dose

Environmental Risk Assessment and Management of Chemicals

DEREK BROWN

1 Introduction

The environmental risks posed by the use of chemicals is an issue of concern to the chemical industry, the general public, non-governmental organizations responsive to public concerns and to the regulator. In the evaluation of such risks a balance needs to be struck between over- and under-protection. That is, a risk assessment and any consequent risk management decisions need to take into account both the potential risks from use and the benefits which chemicals bring.

In any evaluation of risk there is almost always an element of uncertainty in that the database may well pertain only to relatively simple laboratory studies, this certainly being the case for new substances. Even if more detailed laboratory studies and/or field/epidemiological data are available, these may well be limited and less than clear cut. Furthermore, some account may need to be taken of the possible additive effects of more than one chemical substance present in the receiving environment. In this situation, a common method for expressing the uncertainty employs the use of so-called 'safety' or 'assessment' factors to extrapolate laboratory results to a predicted no effect concentration (PNEC) in the field.

Consideration also needs to be given both to the purpose for which the assessment is made and to the nature of the risk under consideration. In terms of purpose, assessment factors may be used to establish water quality criteria (a risk management activity) or as a component of risk assessment. For the former, the quality criteria may be for continuous or intermittent exposure and may also specify the water quality and type to which the discharge is made.

In risk assessment, the PNEC may be considered as one of the two main factors in making an environmental risk assessment, the PEC, or predicted environmental concentration, being the other. The PEC may also be calculated on a conservative basis to reflect uncertainty. In the simplest terms, the risk assessment is made by comparing the PEC with the PNEC.

There are considerable differences in the use to which the result of a risk assessment may be put. It may be of a preliminary nature and used to prioritize work on a large number of chemicals (*e.g.* priority setting in the scope of the EU Regulation on Existing Chemicals[1]). At the other extreme, it may be used as the basis for proposing that a substance poses an unacceptable risk and risk reduction measures are necessary (a possible conclusion from the EU risk assessment of notified substances,[2] or the EU risk assessment of existing chemicals[3]). Thus, depending on the use being made of the results, the nature of the data and the factors employed in the calculation of PEC and PNEC from those data may or may not be critical. If the nature of the risk is such that the effects could be widespread, severe and irreversible, then the level of confidence in the supporting data and the ability of the safety/assessment factors used to predict PNEC and PEC needs to be high. However, if the potential effects are localized and unlikely to be irreversible, then this level of confidence may be reduced.

There also needs to be a clear understanding of what is being protected. In the case of human health the aim is to protect individuals. In the environment, some harm to individual organisms may be accepted, often depending on the emotive response to that organism (housefly *vs.* seal, say). In general, it is accepted that we need to preserve the existence of species and the overall structure and function of ecosystems. Outside of the chemical industry, man's activities, for example in major civil engineering projects, have a major impact on the local ecosystem function and, within constraints, this is largely accepted. This is not suggesting a licence to pollute; rather, that safety factors built into the risk assessment process should provide a realistic balance between over- and under-protection. In the environmental risk assessment of chemicals, this balance critically depends on the appropriate use and selection of assessment factors. Also, these assessment factors need to be considered alongside the PEC exposure scenario in terms of quantities of substance released to the environment and the resulting concentration space/time profile. In other words, is the PEC of short- or long-term duration and is it localized or widespread?

This article addresses the above issues primarily in the context of European and European Union (EU) legislation concerned with the environmental risk assessment and management of chemicals as they are placed on the market. However, trade in chemicals is international and it is important that the different legislations across the world seek to agree a similar approach to their control. Thus, the thinking behind EU legislation has been influenced by and has had influence on international discussions and the legislation in other countries. As appropriate, some reference will be made to this international dimension.

2 Detergent Surfactant Biodegradability

Probably the first environmental management problem to be connected with a

[1] Council Regulation (EEC) No 793/93, *Evaluation and Control of Risks of Existing Substances, Official Journal*, L84/1, 1993.
[2] Commission Directive 93/67/EEC, *Principles for assessment of Risks to Man and the Environment of Substances Notified in Accordance with Council Directive 67/548/EEC, Official Journal*, L227/9, 1993.
[3] Commission Regulation (EC) No 1488/94, *Principles for Assessment of Risks to Man and the Environment of Existing Substances in Accordance with Council Regulation* (EEC) No 793/93, *Official Journal*, L161/3, 1994.

specific class of chemicals, as opposed to problems associated with general waste disposal, arose from the introduction of synthetic surface active agents for use in detergent formulations. Although this issue is now largely historical, the approach taken is of interest in the main context of this article. The questions raised introduced 'environmental issues' to the general public and have almost certainly had a major influence on the way in which subsequent legislation and test methods for assessing biodegradability and aquatic toxicity have developed.

Prior to the 1950s, soap was widely used for laundry purposes. Soap as a detergent surfactant suffers from the major drawback that, as a long alkyl chain carboxylic acid, it forms water-insoluble salts (scum) with calcium and magnesium ions present as 'hardness' in natural water. The first synthetic surfactant to be introduced as a replacement for soap in detergents was tetrapropylenebenzenesulfonate (TPBS), also known as alkylbenzenesulfonate (ABS). This material offered huge advantage in washing performance as it does not form insoluble salts with bivalent metal ions and hence the scum problems associated with soap were not found.

The property of soap in precipitating with hardness salts, together with the high level of biodegradability associated with a linear alkyl chain carboxylic acid, meant that it was very well removed by conventional sewage treatment. By contrast, the TPBS does not form insoluble salts and, as we now know, its highly branched alkyl chain makes it poorly biodegradable. This combination of properties resulted in substantial quantities of undegraded TPBS reaching natural waters with all too evident foaming problems in rivers receiving sewage effluents.

The solution to this particular problem was, with today's knowledge, relatively straightforward and involved the development of biodegradable surfactants. The main anionic surfactant was a straight chain version of TPBS known as linear alkylbenzenesulfonate (LABS). However, in terms of the 1950s and 1960s when the problem became manifest, the solution required a hugh investment of research and effort to develop suitable alternative surfactants, methods for assessing biodegradability, an understanding of the mechanisms of biodegradation and, for the legislator, appropriate laws to address the problem.[4]

The international work on the development and standardization of the methods for the assessment of detergent surfactant biodegradability took place under the auspices of the Environment Directorate of the OECD (Organization for Economic Cooperation and Development). Essentially, these biodegradability methods focus on the primary biodegradation of the surfactant (loss of analytical response) and take no account of whether the molecule is being mineralized (ultimate biodegradation). Furthermore, the reference, or confirmatory, method developed attempts to simulate the operation of a sewage treatment plant. 'Biodegradability' in the context of this method is essentially defined as removal of the analytical response of the surfactant from the aqueous phase. That is, it encompasses both biodegradation and also any removal by sorption onto the activated sludge of the test system.

[4] *Biodegradability of Surfactants*, ed D. R Karsa and M. R. Porter, Blackie, 1995.

Although these biodegradability tests for detergent surfactants, as defined by the OECD,[5] are still current and incorporated into two EU Directives,[6,7] they take a rather simplistic view of environmental risk assessment and management, namely that removal of the analytical response from the aqueous phase during sewage treatment is a sufficient requirement to place high tonnage chemicals on the market.

The world surfactant consumption is 7 Mtonne as surfactant and 8 Mtonne as soap[8] and the surfactant and detergent industry itself goes very considerably beyond the requirements of the EU Directives in assessing the environmental safety of its products, for example as outlined by De Henau *et al.*[9] The EU is currently examining how to update the Detergent Surfactant Directives. Some of the issues involved have been discussed by Brown,[10] who concluded that these Directives have been successful in controlling what is essentially an historic problem and that they continue to have some current utility. Brown also noted that other much newer legislation (*e.g.* The Existing Chemicals Regulation[1]) could be used to investigate whether the current use of specific surfactants poses broader environmental problems, particularly in the context of ultimate biodegradability.

3 OECD/ISO/EU Test Methods

Test methods are a vital component of the environmental risk assessment of chemicals. They need to address end points of concern and as far as possible be predictive of what will happen in the environment, rather than simply be relative measures of the behaviour of chemicals in laboratory studies. Their scope and relevance are one of the major factors in determining the scope and relevance of any subsequent risk assessment.

Stemming from their involvement in the development of test methods for the assessment of the biodegradability of detergent surfactants, the OECD have an on-going Chemicals Programmes. Under this programme, numerous methods for the determination of relevant environmental data for specific chemicals have been developed as OECD Test Guidelines. These continue to be updated and expanded in scope[11] and cover physico-chemical data, degradation data and acute and chronic toxicity data. Many of these have either been adopted or used as EU test methods. Some of this work has also been carried out in collaboration with the International Standards Organization (ISO) whose test methods have also been adopted, or referenced within EU methods. The official EU test

[5] *Proposed Method for the Determination of the Biodegradability of Surfactants*, OECD, Paris, 1976.

[6] Council Directive 82/242/EEC, amending Directive 73/405/EEC, *Approximation of the Laws of the Member States Relating to Methods of Testing the Biodegradability of Nonionic Surfactants, Official Journal*, L109/1, 1982.

[7] Council Directive 82/243/EEC, amending Directive 73/405/EEC, *Approximation of the Laws of the Member States Relating to Methods of Testing the Biodegradability of Anionic Surfactants, Official Journal*, L109/18, 1982.

[8] H. J. Richtler and J. Knaut, in *CESIO 2nd World Surfactant Congress*, 1988, vol. 1, p. 3.

[9] H. De Henau, C. M. Lee and P. A. Gilbert, *Tenside Detergents*, 1986, **23**, 267.

[10] D. Brown, in *Biodegradability of Surfactants*, ed. D. R. Karsa and M. R. Porter, Blackie, 1995, pp. 1–27.

[11] *OECD Guidelines for Testing of Chemicals*, OECD, Paris, 1981; 8th addendum 1996.

methods have been collated by the European Commission.[12] The suite of EU environmental methods in general covers only the 'base set' and, with the exception of the earthworm and inherent biodegradability tests, no EU methods have been published for the studies requested for 'level 1' and 'level 2'. The terms 'base set', 'level 1' and 'level 2' relate to the tests required for the notification of new substances in the EU and will be discussed later.

Physico-chemical Methods

The OECD test guidelines include methods for the determination of water solubility (Guideline 105), vapour pressure (Guideline 104), octanol/water partition coefficient by either the shake flask (Guideline 107) or the HPLC correlation (Guideline 117) methods and adsorption/desorption from soil (Guideline 106). These end points effectively govern the partitioning behaviour of a chemical between air, water, soils/sediments and biota and are thus crucial in focusing on the environmental compartments most likely to be at risk from a substance. Methods for the determination of the water solubility or octanol/water partition coefficient which involve high input of physical energy *via* shaking have been found to give unreliable results for certain classes of sparingly soluble substances.[13]

The physico-chemical guidelines also include a method for determining hydrolytic stability as an indicator of abiotic degradation (Guideline 111). Photolysis in water and in air may also be an important abiotic degradation mechanism.

Biodegradation Methods

The biodegradation test guidelines adopt a somewhat different philosophy to the OECD detergent surfactant biodegradability methods. They use, as analytical end points, non-substance specific measurements such as carbon dioxide evolution, biochemical oxygen demand (BOD) or loss of dissolved organic carbon (DOC) from solution. These end points are directed towards assessing the level of mineralization of the molecule ('ultimate' degradation), rather than, as with the surfactant methods, loss of parent substance.

The OECD biodegradability test guidelines are grouped in three broad classes of 'ready', 'inherent' and 'simulation' tests. The aim of the 'ready' biodegradability tests (Guidelines 301 A–F) is to identify those substances which under non-forcing biodegradability conditions will aerobically biodegrade both rapidly and with a high level of mineralization (ultimate biodegradation). Many substances which do not meet the OECD (and now the EU) definition of 'ready' biodegradation are in common English usage 'readily biodegradable' in practical experience. Blok *et al.*[14] have reviewed many of the issues relating to 'ready' biodegradability studies.

The 'inherent' OECD Guidelines (Guidelines 302 A–C) have been developed

[12] European Commission, *Classification, Packaging and Labelling of Dangerous Substances in the European Union. Part II—Testing Methods*, European Commission, Brussels, 1997.

[13] D. Brown, R. S. Thompson, K. M. Stewart, C. P. Croudace and E. Gillings, *Chemosphere*, 1996, **32**, 2177.

[14] J. Blok, A. de Morsier, P. Gerike, L. Reynolds and H. Wellens, *Chemosphere*, 1985, **14**, 1805.

with the quite opposite philosophy. They use very forcing test conditions and rather relaxed criteria for the rate and extent of biodegradation, with the aim of identifying substances with the potential to degrade. Substances which do not pass an 'inherent' test are unlikely to be aerobically biodegradable.

The 'simulation' test (Guideline 303 A) is designed to simulate sewage treatment conditions and to see to what extent the test substance is likely to be removed from the aqueous phase during conventional sewage treatment. As with the 'ready' and 'inherent' test, the main analytical end point of the simulation test is designed to measure ultimate rather than primary biodegradation. Owing to the test design, the end point for the simulation test relies on a measure of the DOC in the effluent from a test unit in comparison with a control. From the direct experience of the author, these simulation studies, with a DOC endpoint, are extremely difficult to carry out to give reproducible results.

There is also an OECD Guideline (304 A) for biodegradability in soil which relies on the evolution of ^{14}C-labelled carbon dioxide from a ^{14}C-labelled sample of the substance. The Guideline (306), for biodegradability in seawater, is noted in the Guideline to be limited in scope, unless a ^{14}C-labelled sample of the substance is available.

The OECD Test Guidelines for assessing ready biodegradability have, in general, been adopted by the EU for the assessment of biodegradation at the 'base set' level. The inherent and simulation biodegradation methods are used for 'level 1' and 'level 2'.

Apart from the possible experimental difficulties, highlighted above for the simulation study, the approach of concentrating attention on an ultimate degradation end point does have a number of drawbacks for the risk assessment process. The analytical end points of BOD or DOC are not very sensitive and also tend to suffer from a lack of precision. To overcome this, studies using these end points have to be carried out at test concentrations usually in the $10–100\,mg\,l^{-1}$ range. This is extremely high relative to observed environmental concentrations of most xenobiotic substances. Together with other issues, this can make extrapolation of the data to 'real world' biodegradation rates rather problematic.[15-17] Although the lack of sensitivity and precision with the BOD or DOC end points can be overcome by the use of ^{14}C-labelled materials, the toxicity data available tends to be associated with the parent substance. Thus, direct correlation with environmental monitoring experience for the parent substance is not possible.

The above paragraph is not suggesting that the measurement of ultimate biodegradation is without value, since clearly it is important to identify whether degradation is likely to produce stable and possibly problem metabolites. However, it is equally important to assess the loss of parent substance (primary degradation). Arguably, in a simulation test, either of a sewage treatment or a natural water (river) type, primary degradation, normally the parameter most amenable to practical determination, should be seen as the main end point, *i.e.* the end point in the current detergent surfactant Directives.

[15] P. H. Howard, *Environ. Toxicol. Chem.*, 1985, **4**, 129.
[16] M. Alexander, *Science*, 1981, **211**, 132.
[17] M. Alexander, *Environ. Sci. Technol.*, 1985, **18**, 105.

Bacterial Toxicity Methods

The OECD method 209 is based on a procedure[18] designed to identify substances which, if discharged to a sewage treatment works, might inhibit the activated sludge aerobic treatment stage. Since the concentration of activated sludge used in the OECD 209 method is relatively high ($1.6\,g\,l^{-1}$ as dry solids), substances whose toxic effect to bacteria is ameliorated by sorption processes may not show as particularly toxic in this type of study. For the pragmatic purpose for which this study was designed, this relative lack of sensitivity is not necessarily a problem. However, as well as bio-oxidation of organic substances, a well functioning sewage treatment plant should also effect the oxidation of ammonia to nitrate (nitrification) and the inhibition of this important function may not be detected by the OECD 209 method. Nitrification inhibition may be directly determined by exposing nitrifying bacteria to the test substance in a defined medium containing ammonia and measuring the levels of ammonia and nitrate at the end of a specified time period.[19]

Bacterial toxicity study results may be needed to assess potential effects in natural waters or soils. Also, to indicate whether a substance, which shows lack of biodegradation in a ready biodegradability study, could be toxic to the degrading bacteria at the test concentration used.[20] Accordingly, other types of study have been developed in which the effect of the test substance on a relatively low concentration of a bacterial inoculum is measured. For example, growth inhibition may be monitored by turbidity[21] or the effect on the light emission of luminescent bacteria may be measured.[22]

Aquatic Toxicity Methods

The general approach adopted by the OECD and the EU for aquatic toxicity testing is essentially to cover three trophic levels in water, namely algae as primary photosynthetic producers, *Daphnia* as representative of crustacea, and fish as vertebrate animals of food, sport and aesthetic importance to man. This approach assumes that extrapolation, from results obtained for a particular substance on these three trophic levels, will enable a PNEC to be estimated which will be protective of all aquatic species whose exposure is *via* water.

The OECD Guidelines and EU methods for aquatic toxicity testing are essentially short time (conventionally 72 h for algae, 48 h for *Daphnia* and 96 h for fish) with 50% end points being extrapolated statistically from the results so obtained. These three end points form part of both the OECD MPD (minimum package of data) and the EU 'base set' of data. This latter is required for the

[18] D. Brown, H. R. Hitz and L. Schaefer, *Chemosphere*, 1981, **10**, 245.

[19] ISO 9509, *Method for Assessing the Inhibition of Nitrification of Activated Sludge Organisms by Chemicals and Waste Waters*, ISO/TC 147, 1989.

[20] L. Reynolds, J. Blok, A. de Morsier, P. Gerike H. Wellens and W. Bontinck, *Chemosphere*, 1987, **16**, 2259.

[21] ISO 10712, Pseudomonas putida *Growth Inhibition Test*, ISO/TC 147, 1995.

[22] ISO DIS 11348-1/2/3, *Determination of the Inhibitory Effect of Water Samples on the Light Emission of* Vibrio fischeri (*Luminescent Bacteria Test*), ISO/TC 147, 1996.

notification of new substances under the 7th Amendment of the Classification, Packaging and Labelling of Dangerous Substances Directive.[23]

The algae tests (OECD Guideline 201) use either biomass (cell numbers) or growth rates as the measured end points. The latter end point, following recent OECD discussions on the harmonization of classification, is now likely to be preferred for environmental classification purposes. The algal study is multi-generation over the 72 h time frame and should, in principle, be treated as a chronic study. However, the accommodation reached is that the 72 h 50% effect concentration (EC_{50}) is taken as an acute value for the purposes of risk assessment.

The *Daphnia* (OECD Guideline 202) 48 h EC_{50} and fish (OECD Guideline 203) 96 h 50% lethal concentration (LC_{50}) are based on rather simpler end points, namely, immobilization of the *Daphnia* and death of the fish.

For the chronic studies on algae the end point is simply the no observed effect concentration (NOEC) as determined by an identical protocol (often the same study) as that for a 72 h EC_{50} study.

For the *Daphnia*, the chronic study (also OECD Guideline 202) is carried out over 21 days and the main end point is the number of young produced in that time period from parent *Daphnia* which are less than 24 h old at the start of the study.

Two chronic studies on fish have been adopted by the OECD. In the 14 day prolonged fish test (OECD Guideline 204) the end points recorded are fish death and behaviour different from that in the controls. The OECD Guideline 210 is an early life stage toxicity test in which fertilized eggs are allowed to hatch and the fish grown to the stage where they are free feeding. Other fish chronic study protocols are also available for examining the effect of chemicals on growth or the effect on the whole life cycle.

It should also be noted that the proposed update of OECD Guideline 305 (adopted by the OECD on 14 June 1996) for the bioaccumulation of a substance in fish is a flow-through test lasting up to 28 days, followed by a depuration phase in clean water. As well as providing information on bioaccumulation, this method shows whether the concentrations of test substance used have any adverse effects on the fish over the test period.

Soil, Sediment and Avian Toxicity Test Methods

As indicated above, the OECD and many other agencies involved in environmental test method development for general chemicals (as opposed to pesticides) have been somewhat biased towards addressing the aquatic environment. The OECD have Guidelines 207, for the testing of earthworms, 208 for plant growth, 205 for bird toxicity and 206 for effects on reproduction of birds. The need for these tests, especially on birds, should be taken on a case-by-case basis driven by the probability of significant exposure.

Within the EU, there are no officially recognized test methods to assess toxicity to sediment organisms. These organisms have an important environmental role and many substances with a high octanol/water coefficient or cationic nature will

[23] Council Directive 92/32/EEC, amending for the seventh time Directive 67/548/EEC, *Approximation of the Laws, Regulations and Administrative Provisions Relating to the Classification, Packaging and Labelling of Dangerous Substances, Official Journal,* L154/1, 1992.

tend to reach sediments if they are released to the environment. The Society of Environmental Toxicology and Chemistry (SETAC), as a result of a workshop, has issued a guidance document on sediment toxicity tests.[24] Brown *et al.*[13] have described the use of one such test to investigate the toxicity of two phthalate esters to the midge *Chironomus riparius*, exposed *via* spiked sediments.

The OECD Test Guidelines (400 series) on mammals have been designed to extrapolate the results to possible effects on human health. In principle, these data may be used to indicate possible effects on wild populations. By and large this is not done, although the EU Technical Guidance Document[25] (TGD) indicates the principle that secondary poisoning effects on higher trophic levels (fish-eating birds and mammals) should be considered for bioaccumulative and persistent substances.

Limitations of OECD and Other Test Methods

A number of general comments may be made on the OECD and ISO test methods for the generation of toxicity data for the purposes of environmental risk assessment. These comments also apply to the corresponding EU test methods which are, in general, based on the OECD and/or ISO test procedures.

Many of the OECD Aquatic Toxicity (and other) Guidelines, probably because of the detergent surfactant background, seem to have been developed with relatively stable, water soluble, non-volatile, organic substances in mind. It is somewhat ironic that current EU legislation often results in attempts to 'force fit' these methods to substances outside these categories. This can result in very difficult and time-consuming experimental procedures[26,27] to achieve results whose real relevance to environmental risk assessment may, in some cases, be marginal. If a substance is not present in water for any length of time in a bioavailable form, there seems little point in trying to maintain test concentrations in water for a sufficient length of time to carry out aquatic toxicity testing.

The OECD and corresponding EU methods are all carried out in 'clean water'; that is, water containing only inorganic nutrients and buffers. Such studies are of undoubted value for the determination of the inherent toxicity of the substance. They shed little light on the effect that natural water components, such as suspended solids, and dissolved organic substances, such as humic acids, may have on the bioavailability of a substance under investigation[28]. The US EPA[29]

---

[24] SETAC-Europe, *Guidance Document on Sediment Toxicity Tests and Bioassays for Freshwater and Marine Environments*, ed. R. Hill, P. Matthiessen and F. Heimbach, SETAC-Europe, 1993.

[25] Technical Guidance Document in support of Commission Directive 93/67/EEC, *Risk Assessment for New Substances*, and the Commission Regulation (EC) 1488/94, *Risk Assessment for Existing Substances*, European Commission, Brussels, 1996.

[26] P. Whitehouse and M. Mallet, *Aquatic Toxicity Testing for Notification of New Substances: an Advisory Document Dealing with Difficult Substances*, Chemicals Assessment Bureau, Dept. of the Environment, London, 1992.

[27] ECETOC, *Aquatic Toxicity Testing of Sparingly Soluble, Volatile and Unstable Substances*, Monograph No. 26, ECETOC, 1996.

[28] ECETOC, *Environmental Hazard Assessment of Substances*, Technical Report No. 51, ECETOC, 1993.

[29] US EPA, *Fish Acute Toxicity Mitigated by Humic Acid*, Ecological Test Guidelines, US EPA, OPPTS 850.1085, 1993.

have a guideline whereby the mitigating effects on fish toxicity of humic acid may be investigated. The current regulatory approach in Europe is generally to suggest that such effects can be allowed for by considering the physical chemistry of the substance and adjusting the PEC factor accordingly. In the context of risk assessment, there would seem to be considerable value in developing standardized, supplementary test methods using a biological end point to show whether or not normal components of natural water ameliorate the toxicity of a test substance. Similarly, where primary degradation (biotic or abiotic) is known to proceed relatively rapidly, it is possible to devise relatively simple test procedures whereby the degradation is allowed to take place before the test organisms are introduced. This type of procedure has been described for surfactants[30–33] but has gained no official recognition.

Aquatic toxicity test designs generally focus on gross function effects such as death, growth, behaviour and reproductive success. In contrast to mammalian toxicology practice, there has been only limited investigation of biochemical end points and no standardized test procedures are available. There are current concerns that anthropogenic chemicals may be responsible for endocrine modulation effects in the wild, *e.g.* the work of Harries *et al.*[34] on the feminization of male fish as measured by the production of the egg yolk protein vitellogenin. Such concerns may well lead to the development and adoption of standardized tests with an appropriate biochemical end point. However, where a biochemical end point is used, there needs to be assurance that biochemical changes observed in the laboratory are ecologically significant in terms of a wild population.

For rather obvious experimental reasons, most test methods use single species and single substances. In the environment at large there are mixed populations and potential exposure to a mixture of substances. Test protocols have been developed to allow the testing of chemicals to mixed populations, although the test protocols can be complex and the results difficult to interpret. The European Centre for Ecotoxicology and Toxicology of Chemicals (ECETOC)[28] comment that the literature on the effects of mixtures on aquatic toxicity is conflicting. Some authors claim additivity and some do not.

This discussion of some of the limitations of the OECD and related ecological test methods is not designed to be critical of those methods, which perform a very useful function in the assessment of chemicals and their possible effects on the environment. Rather, as mentioned in the introductory paragraph to this section, an understanding of the scope and relevance of the test methods is necessary to understand the scope and relevance of a subsequent risk assessment using those methods.

[30] D. Brown, in *Proceedings of the VIIth International Congress on Surface Active Substances*, Moscow, 1976.

[31] B. Reiff, in *Proceedings of the VIIth International Congress on Surface Active Substances*, Moscow, 1976.

[32] A. Neufahrt, K. Hofmann and G. Tauber, *Spanish Congress on Detergents and Surfactants*, Barcelona, 1987.

[33] R. A. Kimerle and R. D. Swisher, *Water Res.*, 1977, **11**, 31.

[34] J. Harries, D. Sheahan, S. Jobling, P. Matthiessen, P. Neall, J. Sumpter, T. Taylor and N. Zaman, *Environ. Toxicol. Chem.*, 1997, **16**, 534.

4 OECD/EU Environmental Risk Assessment

EU legislation specifically requires that risk assessments (including environmental) are carried out on all newly notified substances and on priority existing chemicals. The OECD has a complementary programme on existing chemicals for the hazard assessment of HPV (High Production Volume) chemicals.

The background legislation which governs the EU risk assessment is summarized by Vosser[35] and by Brown[36] and comes from the 7th Amendment of the Classification, Packaging and Labelling of Dangerous Substances Directive[23] and from the Existing Chemicals Regulation.[1]

The 'Dangerous Substances Directive' dates from 1967,[37] but its scope has been considerably expanded by amendments to the main text and 'adaptations to technical progress' of the annexes to the Directive. The 6th Amendment to the Directive[38] introduced the requirement that 'new' (to the EU market) substances must be notified before they could be placed on the market within the EU. This notification included the requirement for a series of prescribed studies aimed at identifying intrinsic, physical, human health and environmental hazards. These studies are designated 'base set', 'level 1' and 'level 2' which are of increasing complexity and required according to prescribed tonnage triggers. The 7th Amendment to the Dangerous Substances Directive[23] added the requirement that newly notified substances should also be subject to a formal risk assessment. The framework for this assessment in terms of human health (both toxicity risk and risk from physico-chemical properties) and the environment is laid down in a Principles of Risk Assessment Directive[2] and for existing substances in the equivalent Regulation.[3] With the introduction of risk assessment for new substances it seems logical that this, rather than a tonnage trigger, should provide the main rationale for carrying out 'level 1' and 'level 2' studies.

It is within the Annex III to the Risk Assessment Directive[2] and Regulation[3] that the procedures to be used for Environmental Risk Assessment are outlined. There are essentially three stages. In the first, the predicted no effect concentration (PNEC) is calculated. In the second, the predicted environmental concentration (PEC) is estimated. In the third, the risk characterization stage, the PEC/PNEC ratio is used to define what, if any, further action is necessary.

The basic philosophy of the EU procedure is initially to evaluate the PEC/PNEC ratio using very simple data and conservative assumptions. If these data and assumptions lead to the conclusion that there may be a problem (PEC/PNEC > 1), then refinement of either PEC and or PNEC may be carried out. This refinement could well be in several stages using progressively more elaborate data and correspondingly less conservative assumptions. Finally, a conclusion is reached

[35] J. Vosser, in *Environmental Impact of Chemicals: Assessment and Control*, ed. M. Quint, D. Taylor and R. Purchase, The Royal Society of Chemistry, Cambridge, 1996.

[36] D. Brown, in *Environmental Impact of Chemicals: Assessment and Control*, ed. M. Quint, D. Taylor and R. Purchase, The Royal Society of Chemistry, Cambridge, 1996.

[37] Council Directive 67/548/EEC, *Parent Directive Harmonising the Requirements for the Classification Packaging and Labelling of Dangerous Substances*, *Official Journal*, **10**, 196, 267, 1967.

[38] Council Directive 79/831/EEC, amending for the sixth time Directive 67/548/EEC, *Approximation of the Laws, Regulations and Administrative Provision Relating to the Classification, Packaging and Labelling of Dangerous Substances*, *Official Journal*, L259, 10, 1979.

on the PEC/PNEC ratio against a local and regional exposure scenario. If either the local or, more especially, the regional ratio is greater than one, some form of risk management (exposure reduction) may be deemed necessary.

The detailed procedures recommended for environmental risk assessment within the EU are given within the environmental section of the TGD.[25] The stated intention is that the TGD may easily be changed and updated in the light of experience with its use and with new developments in risk assessment methods.

Aquatic Assessment Factors

Both newly notified substances and priority list substances will have as a minimum the 'base set' of data. Within the TGD, and also within the Risk Assessment Directive and Regulation, it is made clear that the normal assessment factor to be applied to acute toxicity data is 1000. That is, the lowest acute toxicity figure (fish 96 h LC_{50}, *Daphnia* 48 h EC_{50}, algae 72 h EC_{50}) is divided by 1000 to give the PNEC. 1000 has been chosen as the standard assessment factor for acute data to be 'conservative and protective and designed to ensure that substances with the potential to cause adverse effects are identified in the effects assessment'.

The assessment factor of 1000 is to some extent an arbitrary choice. However, its use is intended only for initial assessment and it is designed to cover 'inter- and intra-laboratory variation, inter- and intra-species variation, short-term to long-term toxicity extrapolation, and laboratory data to field impact extrapolation'. In this context, it is difficult to argue that this factor should be substantially reduced, although ECETOC used their Aquatic Hazard Assessment (AHA) database to calculate an equivalent assessment factor of 200.[28]

The TGD indicates that for any given substance there may be an argument for raising or lowering the factor and cites a number of situations when consideration could be given to lowering the 1000 factor. These include the availability of data from other taxonomic groups not in the 'base set', data from a variety of species within the three taxonomic groups of the 'base set', and an intermittent release pattern where the TGD specifies that the normal assessment factor applied to acute data should be reduced to 100. Where evidence from other structurally similar compounds and/or knowledge of the mode of action is available, the TGD indicates that the standard 1000 factor may be raised or lowered as appropriate.

As indicated above, the basic philosophy within the EU risk assessment procedure is to treat the PNEC derived from acute data as an initial worst-case estimate. If the PEC, also derived from worst-case assumptions using data on tonnage, use and 'base set' data on degradability and partitioning behaviour, exceeds the PNEC then the likely regulatory decisions will be to refine the PEC and/or PNEC.

PNEC refinement involves further aquatic toxicity testing using longer term chronic studies. If results from chronic studies from two trophic levels (usually two of fish, *Daphnia* or algae) are available then an application factor of 50 is normally applied to the lower NOEC (No Observed Effect Concentration). If three chronic studies are available then a factor of 10 is normally applied to the lowest NOEC. ECETOC[28] proposed that this assessment factor should be 5.

The PEC/PNEC refinement may, in principle, go through a number of cycles and may also include more sophisticated mesocosm toxicity studies and/or ecological observations in the receiving environment (more relevant to high tonnage existing chemicals). Only if this process is still indicating a ratio greater than one, may risk reduction measures be proposed.

Although it is possible to make some criticism of the precise numeric values chosen for the assessment factors in the above scheme, using progressively lower factors as the toxicity database is expanded seems to be entirely logical. The scheme also focuses on freshwater organisms, although it is probable that the relative sensitivity of marine and freshwater organisms to a toxicant has at least as much to do with the influence of seawater on the bioavailability of the substance as with any inherent difference in sensitivity.

There are, however, two particular classes of compounds for which the overall aquatic risk assessment scheme appears to present specific difficulties. The first such class are those substances which can be made bioavailable and show toxic effects in laboratory test systems, but, in the real environment, are effectively rendered not bioavailable through processes such as sorption onto solids or speciation to non/less-toxic forms. In principle, the risk assessment for this type of substance (which includes a significant number of metal compounds) should take such factors into account within the exposure (PEC) term. In practice, this may not be done effectively, especially where the available analytical methods do not differentiate between the 'bioavailable' and 'non-bioavailable' forms. As already mentioned under the section on test methods, this problem could perhaps be addressed experimentally by the development of additional toxicity test procedures to demonstrate whether or not dissolved organics or suspended material present in the real environment is likely to make the substance less bioavailable. Furthermore, detailed understanding of the chemistry of the substance (*e.g.* the oxidation states and co-ordination chemistry of metal ions) is also necessary for a proper understanding of this bioavailability question.

The second class of substance, for which the TGD aquatic risk assessment scheme may present difficulties, are substances which are very sparingly soluble in water and which do not show acute or chronic effects at the limit of water solubility. A possible regulatory approach is to consider the limit of water solubility as the NOEC, but the strict definition of a NOEC is the concentration immediately below an observable effect. Such an approach, especially if standard safety factors are applied, is likely to give quite unrealistic results. Furthermore, for certain substances, in particular certain metals and metal compounds, water solubility may have no true meaning in that any 'solubility' measured is essentially reaction with the water or its constituents to give soluble metal ions. This issue is currently being considered within the EU and elsewhere for metals and metal compounds in the context of environmental classification.

STP Micro-organism Assessment Factors

The TGD refers to the need to decide the likely PNEC for substances entering a sewage treatment plant from the point of view of inhibiting the microbial activity in the plant. The test methods and recommended assessment factors are:

inhibition of growth of *Pseudomonas putida* or the activity of nitrifying bacteria by ISO 9509 (assessment factor of 1 on the NOEC and of 10 on the EC_{50}); inhibition of respiration rate of activated sludge by OECD 209 (assessment factor of 10 on $NOEC/EC_{10}$; assessment factor of 100 on EC_{50}).

It should be noted that the tests proposed do not cover the possible inhibition of anaerobic bacterial processes. The sludge digestion process is both of major importance to those works which practice it and also rather sensitive to certain substances.

Sediment Assessment Factors

The TGD acknowledges that no data for sediment dwelling organisms will be available for newly notified substances. The TGD also comments that few tests with sediment dwelling organisms have been carried out in Europe and that research into sediment test methodology is still under way. Thus, sediment test data are also unlikely to be available for most existing substances.

The TGD suggests that, in the absence of sediment organism test results, the equilibrium partitioning method should be used for a provisional calculation of the sediment PNEC. This makes the assumption that the sediment dwelling organisms have a similar sensitivity to water column organisms and that the exposure route of the sediment organisms is *via* the interstitial water in equilibrium with the sediment.

Terrestrial Assessment Factors

The TGD recognizes that the terrestrial ecosystem covers organisms other than those found in the soil, but essentially limits the consideration of the terrestrial compartment to soil.

Two approaches are given in the TGD for the calculation of PNEC in the soil. The first follows the equilibrium partitioning method for sediments as described above. Unlike the case of sediment organisms relative to free swimming organisms, where there may be a reasonable presumption of similar response to toxic substances, this correlation between free swimming aquatic organisms and soil organisms seems much less secure. Hence, this approach cannot, on the available evidence, be supported.

The second approach uses assessment factors and exactly parallels the scheme for the aquatic environment. However, the TGD states that information on soil organisms will only be available for some substances and where data are available they are only likely to be on earthworms. Furthermore, 'an internationally accepted set of standardized ecotoxicological tests for hazard assessment of chemicals for the soil compartment is not available for the moment'. Within this context, it seems premature for the TGD to have proposed assessment factors in that there is little or no basis on which to use them in any rational way.

This issue is of particular importance for metals which can reach agricultural land through the use of sewage sludge as an agricultural fertilizer. There are, however, well established guidelines for the acceptable levels of a number of metals in agricultural soil and it would seem sensible to employ these rather than to attempt to calculate a PNEC using the procedures proposed in the TGD. However, the TGD does recognize that metals and metal compounds may not be treated in exactly the same way as organic substances and accordingly gives further guidance in an Appendix VIII as to how environmental risk assessment should be carried out. The appendix on metals make the important comment that 'calculated PNECs derived for essential metals may not be lower than natural background concentrations'.

OECD Assessment Factors

The OECD took a view on assessment factors broadly similar to that in the EU TGD. The main differences are that the OECD use an assessment factor of 1000 applied to a single $L(E)C_{50}$ value and of 100 to the lowest $L(E)C_{50}$ values from a set consisting of at least algae, crustaceans and fish.

The OECD also allow the use of QSAR (Quantitative Structure–Activity Relationship) data to estimate both NOEC chronic toxicity values and acute toxicity $L(E)C_{50}$ values. This approach provides a saving in experimental resource and animal life and, with the caveat that the QSAR used has been properly validated, could usefully be adopted by the EU. The EU discusses the use of (Q)SAR under Chapter 4 of the TGD, but does not currently allow QSAR data to be used in the data sets to calculate PNEC.

As with the EU TGD scheme, the OECD proposes an assessment factor of 10 as applied to the lowest chronic NOEC. The OECD also envisages data from (semi-)field tests to provide a basis for a more comprehensive effects assessment in combination with chronic toxicity data.

Predicted Environmental Concentration (PEC)

As a general principle, it seems self-evident that the need for testing should be driven primarily by exposure considerations and that tests on higher animals should certainly need a strong indication of significant exposure.

The TGD approach is to consider both local and regional PECs using generic EU scenarios. In terms of the local PEC, this approach of a generic local scenario seems unlikely to give realistic PEC results, especially for point source discharges when coupled with standardized emission factors for the various use categories (Appendix 1 of the TGD). Where such standardized calculations of PEC_{local} are made, they should be considered very much as initial estimates and every effort made to refine the calculation based on more site-specific and substance-specific data. It this is not done, the PEC_{local} may be unrealistically high and lead to a large amount of unnecessary testing. This can be a waste of resource, including the lives of the test animals, and also can have a severely inhibiting effect on the development of new and possible safer chemicals.

A recent ECETOC report[39] has reviewed exposure assessment in the EU TGD and also concludes, in relation to both environmental and human exposure, that the current system is over-conservative and likely to lead to unnecessary testing.

---

[39] ECETOC, *Exposure Assessment in the Context of the EU Technical Guidance Documents on Risk Assessment of Substances*, Document No. 35, ECETOC, 1997.

This ECETOC report makes the interesting proposal, to be backed by future work, that 'risk characterization' as a PEC/PNEC ratio should be replaced by 'risk estimation'. That is, expressing the likelihood of the incidence and severity of effects and stated in terms of a statistical distribution about a most probable value and not by a single number.

ECETOC[39] also discusses the rate constants for biodegradation assumed in the EU TGD. In relation to removal in waste water treatment plants and degradation in surface waters, sediments and soils, the rate constants in the TGD are probably extremely conservative. Again, conservatism in terms of degradation rates will almost inevitably lead to over-testing.

The $PEC_{regional}$ calculation is usually significant only for large tonnage substances where significant release to the environment may occur. As ECETOC[39] comments, steady state concentrations calculated by the fugacity approach are usually assumed, although in reality $PEC_{regional}$ may show considerable variation with time. For large tonnage chemicals the resource is usually available to confirm or otherwise the estimates of $PEC_{regional}$. Furthermore, even if these estimates are somewhat conservative, it is arguable that the PNEC of substances which have a widespread distribution in the environment at significant concentrations should be calculated on a reasonably substantial toxicity database.

Bioaccumulation and Secondary Poisoning

The concern that bioaccumulation of persistent and possible toxic chemicals in fish can introduce those chemicals into the human food chain forms the basis of the Japanese MITI (Ministry of International Trade and Industry) laws governing the manufacture and import of new substances. In outline, these laws require that a new substance in Japan is examined by a laid down biodegradability procedure (essentially that described in OECD 301C) and, if not judged to be acceptably biodegradable, by a specified 8 week bioaccumulation study. The use of a substance which is judged to be both not biodegradable and bioaccumulative is banned or severely restricted, unless it can be shown by a series of rigorous mammalian toxicology studies to be safe for humans.

The Japanese concern in respect of persistent and bioaccumulative chemicals can be well understood in the light of their most unfortunate experience with the mercury poisoning incident at Minamata Bay.[40] An industrial release of mercury bioaccumulated in the fish, which formed the main diet of the local population and caused major dysfunction of the central nervous system. However, it is arguable that the Japanese legislation is both too simplistic and too detailed. The legal specification of the precise test methods to be used does not allow for the frequent need to adapt test methods to the type of substance being tested. More fundamentally, the legislation takes no account of the quantities of the substance likely to be released to the environment. The test procedures would also be unlikely to pick up the subtleties of the Minamata Bay incident in which it appears that the mercury discharge was biomethylated in the environment and that methylmercury was the bioaccumulated form.

[40] C. Baird, *Environmental Chemistry*, Freeman, New York, 1995.

In the EU, the TGD approaches the issue of bioaccumulation and secondary poisoning primarily from the viewpoint of protecting fish-eating birds or mammals. The TGD provisional approach proposes that, for substances with a bioaccumulation potential and for which mammalian toxicity tests indicate the potential for harm, the PEC_{oral} calculated from PEC_{water} multiplied by the calculated or measured BCF (bioconcentration factor) should be compared with the $PNEC_{oral}$. This latter is calculated from the available toxicity data for birds and mammals. ECETOC[41] has reported on the role of bioaccumulation in environmental risk assessment and this document forms a useful supplement to the TGD.

Environmental Risk Assessment Conclusions

It is self-evident that the quality of a risk assessment is critically dependent on the quality of the data used to generate the PEC and PNEC factors. In the section on test methods, some of the potential problems with current toxicity and degradation test procedures have been indicated. In indicating these problems it is not being suggested that the risk assessments made on the basis of the test methods are invalid, rather that the results obtained need to be considered very carefully for their relevance to the real world situation.

In particular, the theoretical and/or measured levels of exposure (PEC term) should be considered together with the physical chemistry of the substance in order to evaluate whether the PEC is actually likely to be bioavailable. It has been suggested in the test methods section that additional toxicity test procedures could usefully be developed to shed experimental light on this bioavailability issue. The use of standardized scenarios to calculate PEC_{local} can lead to overly conservative estimates. Real site scenarios should be used wherever possible and a judgement taken as to whether the overall amount of the substance released is likely to pose a significant problem. This issue is particularly important for new substances where unrealistic PEC_{local} calculations could generate unnecessary demands for 'level 1' and 'level 2' toxicity tests at very low total tonnages.

The PNEC term in the risk assessment is generally derived from tests carried out on relatively few organisms. The test substance is generally the sole potentially active substance present and rather simple test end points are used. While recognizing these limitations, it is not appropriate simply to increase assessment (safety) factors to take the limitations into account. If this is done, the factors rapidly become so high that virtually every chemical becomes suspect. Where specific issues, such as the possibility of endocrine modulation effects, are identified, targeted work either to develop a standardized screening test or to investigate specific suspect substances should be initiated.

As with PEC, the results of PNEC calculations should be cross-checked against what is known in the real world situation. If PNEC calculations give levels lower than are known to exist in an apparently satisfactory ecosystem, further investigation of the reasons (*e.g.* incorrect data, bioavailability, acclimation of the environmental species) is required.

[41] ECETOC, *The Role of Bioaccumulation in Environmental Risk Assessment*, Technical Report No. 67, ECETOC, 1995.

Thus, as an overall comment, environmental risk assessment methodology, *e.g.* as exemplified by the EU TGD, is still very much an evolving science and risk management decisions based on the results do need to recognize these limitations.

5 Environmental Risk Management

In the context of chemical substances, whose hazardous properties are intrinsic to the specific substance, environmental risk management measures effectively involve reducing the exposure of the environment. As has been pointed out by Brown,[36] an underlying objective of the Classification, Packaging and Labelling of Dangerous Substances Directive, from its origins[37] to its most recent significant amendment,[23] has been risk management. The Directive requires the user of a substance to be warned of intrinsic hazardous properties *via* appropriate labelling. For professional users this warning label is reinforced with a safety data sheet which gives the information 'necessary for the protection of man and the environment'. Thus, advice on avoiding spillage to drain and on the appropriate methods of treating, recycling or disposing of waste should be provided so that the user can minimize release to the environment.

On this basis, the outcome of the formal risk assessment procedures for new and priority existing substances could include specified labelling, safety data sheet warnings and specified waste treatment systems. Since such risk reduction measures all affect the PEC, they need to be considered as an on-going part of the risk assessment process rather than as a final stage.

Vosser[35] has briefly reviewed the control measures available to individual Member States for a substance it considers harmful and also the Community-wide legislation under the 'Marketing and Use' Directive[42] which is amended as different substances become subject to EU-wide control measures.

There are numerous other national, EU and international agreements which control directly or indirectly the quantities of specific chemicals or chemical classes (*e.g.* pesticides) which may be used in or discharged to the environment and hence are risk management measures. However, this article will only indicate the outline of a few of these.

The EU Directive 76/464[43] is aimed at preventing pollution of the aquatic environment by specified chemical substances or families of chemical substances. Substances for inclusion in this Directive have been selected primarily on the grounds of their persistence, toxicity and bioaccumulation properties. The Directive divides the dangerous substances into two groups. Substances on List 1 (the 'black list') are subject to Community-wide controls and on List 2 (the 'grey list') to control by individual Member States. The Directive allows for controls to be based either on specified effluent discharge levels or on environmental quality objectives (EQOs). Most Member States have opted for control of levels on

---

[42] Council Directive 76/769/EEC, *Framework Directive for Restriction on the Marketing and Use of Dangerous Substances and Preparations*, amended as required as new substances to this Directive are to be subject to control measures across the European Union, *Official Journal*, L398/19, 1976.

[43] Council Directive 76/464/EEC, *Framework Directive on Pollution Caused by Certain Dangerous Substances Discharged into the Aquatic Environment*, *Official Journal*, L129/23, 1976.

effluents and these have primarily been set using the BATNEEC ('best available technique not entailing excessive cost') principle. The UK historically opted for control through the EQO route, but the Environment Protection Act of 1990 effectively requires the application of a combination of both EQO and BATNEEC.

The EQO approach is essentially based on a risk assessment-type consideration, in that the level allowed in the receiving water is set to be below that which would harm the organisms present. The effluent discharge controls, which essentially aim at the reduction of levels to the lowest possible, are based more on the 'precautionary principle' approach.

The 'Precautionary Principle'

As accepted by the Oslo Paris Commission (OSPARCOM), this essentially states that the release of persistent, toxic and bioaccumulative substances to the marine environment should be reduced at source by the use of best available technology. This applies especially when there is reason to assume that damage to living marine resources is likely, even when there is no scientific evidence to provide a causal link.

The precautionary principle philosophy essentially says that for some substances no level is 'safe'. While this may be true for certain substances, such as PCBs and the like, it is a conclusion which should only be reached following a rigorous risk assessment and a strong indication of possible major environmental harm. Criteria based on 'base set' type tests, such as those used for the EU classification 'Dangerous for the Environment' (see Brown[36]), are not appropriate for the application of this principle.

Ozone Depleting Chemicals

The association of the use of volatile substances containing particularly the C–Cl or C–Br moieties with the catalytic destruction of the stratospheric ozone layer is now well established. However, it is nevertheless a complex atmospheric chemistry issue requiring very detailed studies and research[40] to achieve what is, even now, an almost certainly incomplete picture of the ozone destruction process. The importance of the ozone layer in filtering out potentially cancer-inducing UV light has led to international agreements (the 'Montreal Protocol', *etc.*) to phase out the production and use of many ozone-depleting substances. These agreements have focused primarily on the CFCs (chloro-fluorocarbons), which have been identified as being the most stable in the atmosphere and hence having a major ODP (ozone depletion potential).

CFCs are otherwise extremely safe and have found widespread and important use, for example as refrigerants, aerosol propellants, blowing agents for structural foams and as fire fighting agents. Hence the development of suitable alternatives continues to pose major technological challenges. Nevertheless, the international agreements to phase out production and use surely represent an appropriate response. Although the benefits from the CFCs have been large, the magnitude of the danger from their continued use is quite unacceptable.

6 Conclusions and Summary

At the outset of this article, the point was made that the risk assessment process and any consequent risk management decisions need to be well balanced as between the magnitude and severity of the potential risks posed by chemical substances and the benefits to society which those chemicals bring. Where the problem is shown to be potentially globally catastrophic, as in the destruction of the ozone layer, judging the balance becomes essentially self-evident. On a far lesser scale of impact severity, the phasing out of poorly degradable surfactants in detergent formulations provides a good example of how a demonstrable problem led to an appropriate technical solution.

However, much of the current OECD and EU activity in risk assessment has been focused on more local or regional issues where the evidence of damage from specific substances may be missing and the degree of uncertainty as to whether a substance is causing harm may be high. In this situation, any risk management measures are directed more towards preventing possible problems than solving those which have become manifest. Here, achieving a balance between the risk and the benefit is altogether more difficult.

In this type of pro-active risk assessment and management, the test methods used to assess the possible environmental impact of substances are critical in establishing the scope and relevance of the risk assessment/management decisions which may stem from the test method results. Accordingly, some considerable part of this article has been given to a discussion of the scope and limitations of current methods, which aim to provide information on environmental distribution, degradation and ecotoxic properties.

The current aquatic toxicity methods used to establish inherent toxicity are essentially carried out on single species and with gross effect (*e.g.* death, behaviour or reproductive success) end points. These methods could usefully be supplemented by methods using a biological monitor to demonstrate how degradation and sorption processes affect the toxicity of the test substance. Provided that the ecological significance is established, standard methods using more subtle biochemical end points, to monitor for effects such as endocrine modulation, may also be of value.

In respect of risk assessment, the EU TGD approach of a progressive refinement of the PEC and PNEC terms is broadly logical, although a number of detailed concerns are raised. The need for testing, especially on higher animals, should be driven primarily by exposure considerations. PEC_{local} calculations using generic local scenarios are likely to give unrealistically high values and, wherever possible, site-specific and substance-specific data should be used. The limitations of the current test end points need to be recognized. Most importantly, the results obtained using the TGD procedures need to be cross-checked against the real world situation.

The environmental risk management of substances effectively involves exposure reduction and the warnings provided by labelling and appropriate safety data sheet advice on use, disposal and waste treatment can do much to achieve this. Any such reductions need to be factored into PEC calculations as an integral part of the risk assessment process.

Whilst it must be recognized that it is adverse properties of a substance that may make stringent risk management decisions necessary, these decisions should also be based on a reasonable presumption of exposure and significant harm to populations and ecosystems. In the context of the 'precautionary principle', it is suggested that particular attention is given to these caveats. Substances should not be targeted for bans solely on the basis of adverse data obtained from rather simple laboratory studies.

Assessment of Risks to Human Health from Landfilling of Household Wastes

GEV EDULJEE

1 Introduction

Background

The landfilling of municipal solid waste (MSW) and, in particular, household waste is a widely adopted waste management practice; indeed, some 90% of the UK's household waste is disposed of in this manner. In recent years there has been a move to develop waste management solutions based on the concepts of the Best Practicable Environmental Option (BPEO), life cycle assessment and the so-called 'waste management hierarchy' whereby recycling and beneficial use of waste are given priority over ultimate disposal.[1] To this end the Waste Regulation Policy Group of the Environment Agency has been considering the BPEO for particular waste streams as part of a sustainable and integrated approach to waste management in the UK. In most cases there will be a choice of waste management options for planners, operators and regulators to consider, and for household waste this choice will inevitably include landfilling.

All waste management options, including landfilling, involve an element of risk to human health.[2] In seeking to identify the BPEO for household waste, the Environment Agency needs to assess the relative health risks associated with different household waste management options. The present article summarizes the outcome of a project commissioned by the Environment Agency, specifically focusing on the public health risks associated with household waste disposal by landfilling.[3] The article commences with a brief description of the health risk assessment framework, followed by a discussion on each of its elements. A risk assessment methodology is developed by systematically examining the potential releases from household waste landfills. Finally, the application of the risk assessment methodology is illustrated by a case study drawn from actual landfill

[1] G. H. Eduljee and D. A. Arthur, in *Pollution: Causes, Effects and Control*, ed. R. M. Harrison, Royal Society of Chemistry, Cambridge, 3rd edn., 1996.

[2] O. Bridges and J. W. Bridges, *Comparison of the Risks from Landfill and Incinerators of Municipal Solid Wastes*, University of Surrey, Guildford, 1995.

[3] Environment Agency, *An Assessment of the Risks to Human Health from Landfilling of Household Wastes*, Report No. CWM 143/97 and 143A/97, Environment Agency, London, 1997.

siting and operating conditions in the UK. Detailed discussion and calculations can be found in the project report.[3]

Chemical Processes Within Household Waste Landfills

On a dry weight basis the primary constituents of household waste are carbohydrates such as celluloses, starch and soluble sugars (in excess of 40%), ferrous and non-ferrous inorganics (30%) and moisture (20%). In addition to these major components, household waste often contains small quantities of a wide variety of chemical wastes such as pesticides and herbicides from gardens, oil, grease, waste solvents and paints, industrial chemicals, detergents and cleaners, *etc*. These materials introduce trace inorganic and organic chemicals into household waste. Constituents such as heavy metals are not degraded by the chemical processes within the landfill but are nevertheless chemically transformed and generally retained within the site. Other constituents are broken down along with the principal components of household waste and/or are released to the environment along with landfill gas and leachate.

The fate of trace inorganic and organic chemicals in household waste is a function of their physical and chemical properties, and of the reaction conditions within the landfill. Organic chemicals can be degraded by micro-organisms in the landfill and by hydrolysis in groundwater environments. Thus, chlorinated organics can be dehalogenated under anaerobic conditions to produce lower molecular weight chlorinated compounds such as vinyl chloride, which can be released from the site *via* landfill gas or *via* leachate. Some chlorinated aromatics are also amenable to degradation, but fragmentation of the aromatic ring requires the presence of oxygen. While organic species such as dioxins are not readily degraded, they are strongly hydrophobic, adsorbing onto material in the fill and, therefore, remaining highly immobile. Metals are not in themselves degraded; mobilization within the landfill, primarily *via* leachate, depends on the species and reaction conditions. In the aerobic acid-forming stage of a landfill, metal salts can be solubilized in leachate to a greater extent than in the anaerobic stages of the landfill when metal salts are reduced to less soluble forms. Metals also are strongly adsorbed onto fill material.

Risk Assessment Framework

The procedure for assessing the health impacts of household waste landfill sites can readily be accommodated within a general framework for risk management applicable to all waste management operations. The framework comprises the following elements:

- *Hazard identification*: Consideration of the properties and concentrations of the chemicals of interest that could give rise to harm
- *Exposure assessment*: Identification of potentially exposed populations; modelling of emissions, fate and transport of chemicals and other releases through the environment; estimating the dose received by the populations potentially at risk

- *Toxicity assessment*: Definition of the dose–response relationship for each chemical of interest
- *Risk estimation*: Calculation of quantitative estimates of carcinogenic and non-carcinogenic risks to receptors for all the exposure scenarios considered
- *Risk evaluation and control*: Judgements as to the significance and acceptability of the estimated risks; formulation and implementation of decisions about tolerating or altering the risks

The elements comprising hazard identification, exposure assessment, toxicity assessment and risk estimation and evaluation can be grouped under the term *risk assessment*, while the entire process, including risk control, can be termed *risk management*.

2 Hazard Identification

A prerequisite for health risk assessment is a clear understanding of what chemicals are present at a site, their concentration and spatial distribution, and how they could move in the environment from the site to potential receptor points. It is not rare to detect over 100 different chemicals in household waste and its associated releases (landfill gas, leachate and dust). The hazard identification stage examines the data for these contaminants and consolidates the data to stress the chemicals of concern (*i.e.* those representing the majority of risk posed by the site). The surrogate chemicals are selected on the basis of which compounds best represent the risk posed by the site:

- The most toxic, persistent and mobile
- The most prevalent in terms of frequency of detection and concentration
- Those involved in the most significant exposures

The selected indicator chemicals together with their default concentrations are presented in Appendix 1. Their selection is discussed in detail elsewhere.[3]

3 Release and Transport of Contaminants

The risk analyst must next estimate the concentration of contaminants at the exposure points, *via* all relevant pathways—air, ground and surface water, soils/waste and, if appropriate, food. Mathematical models are applied to calculate the transport of chemicals through air and water. For groundwater contaminants, hydrogeologic models can be used to estimate the concentration at a downstream well. For organic chemicals released to the atmosphere, diffusion models can be employed along with representative meteorological conditions to estimate downwind concentrations on and off the site.

Generation and Transport of Landfill Gas

Gas Generation Rate. There are several approaches to the estimation of gas generation rate. For example, the microbial degradation processes can be modelled, taking into account the site water balance, waste density and waste composition. A related approach is to model the rate of acetate metabolism in the leachate.[4] A screening approach that serves the purpose of the risk management framework is either to adopt a single gas generation rate for 'fresh' and 'old' landfills, respectively, or to simulate gas generation rates with respect to time by algorithms fitted to experimental gas generation data. Regarding the first option, the following raw gas production rates have been suggested:[5]

- Waste landfilled for less than 5 years: $15 \, m^3 \, tonne^{-1} \, yr^{-1}$
- Waste landfilled for more than 5 years: $5 \, m^3 \, tonne^{-1} \, yr^{-1}$

Alternative algorithms yield gas generation rates of 6–$15 \, m^3 \, tonne^{-1} \, yr^{-1}$ and 4–$7 \, m^3 \, tonne^{-1} \, yr^{-1}$, respectively, the ranges corresponding to low and high yields.[6] Extensive studies of UK landfills have indicated far lower gas generation rates, in the order of 1–$3 \, m^3 \, tonne^{-1} \, yr^{-1}$ for sites of all ages. Therefore a conservative gas generation rate of $5 \, m^3 \, tonne^{-1} \, yr^{-1}$ was applied to landfills of all ages.

Transport of Landfill Gas. The impact of emissions from a landfill site on off-site receptors can, subject to various assumptions, be assessed with acceptable precision using existing Gaussian dispersion models. This is not the case for receptors on the landfill itself (*e.g.* banksmen), or very near to it (site office workers). Gaussian point source dispersion models, even if modified to allow for finite source dimensions, are inappropriate in assessing exposure of site workers. There are a number for reasons for this, the most important being that Gaussian models break down at low values of source–target separation: as separation tends to zero, modelled concentrations tend to infinity. On-site concentrations emanating from an area source may also be roughly approximated from mass conservation using a box model.[7] Such analyses have produced plausible results, but suffer from the same fundamental defect, namely, reliance on Gaussian formulae in the very near field. An alternative algorithm for on-site landfill gas concentrations has been derived[3] on the basis of field measurements of fine dust particles.[8] The algorithm refers to concentrations at inhalation level (1.6 m above the ground), at the centre of the downwind edge of the active source on site. It therefore gives the highest concentrations to which (above ground) site workers would be exposed.

For off-site receptors, dispersion from area sources is most conveniently treated by positioning an equivalent virtual point source upwind of the area

[4] Department of the Environment, *The Technical Aspects of Controlled Waste Management: Appraisal of Hazards Related to Gas Producing Landfills*, DoE Report No. CWM/016/90, Department of the Environment, London, 1990.

[5] S. Cernuschi and M. Guigliano, in *Proceedings of the First International Landfill Symposium*, Department of the Environment, London, 1989.

[6] C.S. Bateman, in *Proceedings of the Fourth International Landfill Symposium*, Department of the Environment, London, 1993.

[7] US EPA, *A Workbook of Screening Techniques for Assessing Impacts of Toxic Air Pollutants*, EPA-450/4-88-009, Office of Air Quality, Planning and Standards, Research Triangle Park, NC, 1988.

[8] J. H. Shinn, N. C. Kennedy, J. S. Koval, B. R. Clegg and W. M. Porch, in *CONF-740921 ERDA Symposium Series 38*, Energy Research and Development Association, Oak Ridge, TN, 1976, pp. 625–636.

source. The concentration resulting from the point source is then calculated using a standard Gaussian formula. An alternative approach, adopted in this risk assessment,[3] is to extend the model for on-site transport of landfill gas.

Release of Combusted Gas. In order to develop a source term for combusted gas emissions, we assume that combustion of landfill gas will result in $10\,m^3$ of flue gas per m^3 of raw gas (flue gas at 11% oxygen). The transport of emissions from a landfill gas flare or gas engine will behave differently from fugitive releases of landfill gas. Flares or gas engines are point sources and have thermal buoyancy and/or momentum, as a result of the temperature and velocity of the gases emitted from the stack. Therefore, it is not possible to apply the methodology for landfill gas releases to emissions from flares or gas engines. For flares, the calculation procedure developed by the US EPA[7] was applied. The calculations provide the assessor with an estimate of the emission rate of a release, and the height from which the release is effected. The effective stack height for emissions from the stack of a gas engine can be calculated from a standard Gaussian treatment of stack emissions.[9]

Ground Level Concentration of Emissions. The dilution of the emissions from the point of exit from the stack to ground level will be dependent on a number of factors, particularly:

- Size of the source
- Wind speed
- Atmospheric stability
- Stack height
- Temperature of the gases

For the purposes of a screening assessment, a simplified approach was adopted whereby the calculation of ground level concentration is limited to the maximum concentration of a pollutant along the centre line of the plume; this represents the maximum pollutant concentration at ground level following release of that pollutant from the stack. The concentration at ground level of a pollutant along the centre line of the plume was calculated using algorithms giving the characteristic Gaussian distribution of ground level concentrations along the centre line, with a maximum occurring some distance away from the stack.[9]

Generation and Transport of Dust

Dust can be generated by the following mechanisms:

- Erosion of the landfill surface by the action of wind
- Erosion of the landfill surface by vehicles driving on the site
- Generation of dust from landfill operations such as discharging waste from tipper trucks and the spreading of waste and daily cover material

[9] M. Smith, *Recommended Guide for the Prediction of the Dispersion of Airborne Effluents*, American Society of Mechanical Engineering, Washington, DC, 1968.

The potential for an adverse health effect following exposure to dust is dependent on whether the generated material is contaminated, and on the concentration of the dust at the point of exposure. If the surface of the landfill is composed of clean cover material, then the dust generated and transported to the point of exposure will also be 'clean' from the perspective of generating an adverse health effect following inhalation or dermal contact.

Dust Generation by Wind Erosion. Erosion of the surface of a landfill by wind generates dust particles which can be transported off site, depending on the strength of the wind. Of interest from a health standpoint is the concentration of particulate matter (PM) that is below $10\,\mu$m in diameter (so-called PM_{10} particles), owing to their ability to penetrate into and be retained in the lungs. The emission factor for PM_{10} erodible particles is a function of the dust flux rate (in units of $g\,m^{-2}\,d^{-1}$), the mean wind speed (taken as $5\,m\,s^{-1}$), the threshold wind speed (taken as $7.5\,m\,s^{-1}$) and the fraction of vegetation cover (taken as 0.5 for active sites and 0.9 for closed sites).[7] The rate of emission of a chemical in the dust is obtained by multiplying the concentration of the chemical in soil or dust by the site area over which erosion occurs.

Dust Generation from Landfill Operations. Landfill operations which have the potential to generate dust are the offloading of waste from the transporter, spreading of waste and of daily cover material, and vehicular movements on the landfill surface. The algorithms proposed[7] account for these sources of dust generation. The key parameters influencing dust generation are percentage of silt and moisture content of the waste or road surface; the mean vehicle speed; mean vehicle weight and number of wheels; mean wind speed; and the number of days with at least 0.25 mm of precipitation per year. The emission rate of the chemical associated with the waste/dust is the product of the concentration in soil or dust and the emission factor, in compatible units. The emission occurs over the duration of an event such as offloading, or during the time a vehicle traverses the site.

Atmospheric Transport of Dust. For site workers in the immediate vicinity of dust emissions, a simple box model was used, setting the height of the box at 2 m to accommodate the breathing height of a site operator, while the width of the box was represented by the equivalent side length of the working area (for example, the working face). The ambient air concentration of contamination within the dust medium can be obtained by multiplying the ambient air concentration of dust/soil by the concentration in dust/soil of the contaminant.

For site workers removed from the point of generation of PM_{10} dust, and for off-site populations, a consistent approach was taken by adopting the algorithm representing the transport of landfill gas.[3] In this case the source term was averaged across the dimensions of the site.

Generation and Transport of Leachate

Landfill leachate can be produced as a result of rainfall infiltrating the site, and as a result of the biodegradation reactions within the landfill. Leachate can move downwards through the base of the landfill and into an underlying aquifer, or it can rise to the surface as seepage and drain to a surface water body such as a stream or a reservoir. The leachate generation rate varies with time, depending at any one instant on a balance between ingress of water into the landfill, the removal of leachate from the landfill by pumping or by seepage, and the permeability of the base material. After release from the landfill/liner system, the leachate moves through the unsaturated layer of soil and then into the saturated zone or aquifer. Groundwater provides a medium of transport for the leachate from the point of generation to a location downgradient of the site where a receptor could potentially be exposed *via* ingestion or dermal contact. Surface waters impacted by leachate or groundwater contaminated by leachate could also form a potential exposure medium either directly (ingestion, dermal contact) or indirectly (*via* consumption of fish).

In the present risk assessment the attenuation of inorganic and organic chemicals as the leachate passes vertically through the unsaturated zone beneath the landfill and the subsequent lateral transport of leachate *via* groundwater to an offsite receptor has been modelled by LANDSIM,[10] a bespoke computer package which uses a Monte Carlo simulation technique to sample randomly the model variables within a defined probability distribution.

Leachate can also travel as seeps to surface water ditches and ponds. The exposure point concentration is a function of the seep flow rates and concentrations, and the dilution of the leachate in the receiving waters. The seep flow should ideally be determined through field observations or calculated using a detailed landfill water balance. For screening purposes the concentration of the chemical in the receiving water body can be pro-rated according to the flow rate of the seep relative to the flow rate of the surface water body.

4 Exposure Assessment

Exposure Pathways

A chain of events must occur to result in exposure. The chain in a collective sense is termed an exposure pathway, the environmental routes by which chemicals from the site can reach receptors. A pathway defines the framework of a fate and transport analysis and consists of the following elements:

- A source
- Chemical release mechanisms (*e.g.* leaching)
- Transport mechanisms (*e.g.* groundwater flow)
- Transfer mechanisms (*e.g.* sorption)
- Transformation mechanisms (*e.g.* biodegradation)
- Exposure point (*e.g.* residential well)
- Receptors (*e.g.* residents consuming potable water from the well)
- Exposure routes (*e.g.* ingestion of water)

The exposure pathways considered in this study are illustrated in Figure 1.

[10] Golder Associates, *Landsim Manual Release 1*, Reference No. CWM 094/96, Golder Associates, Nottingham, 1996.

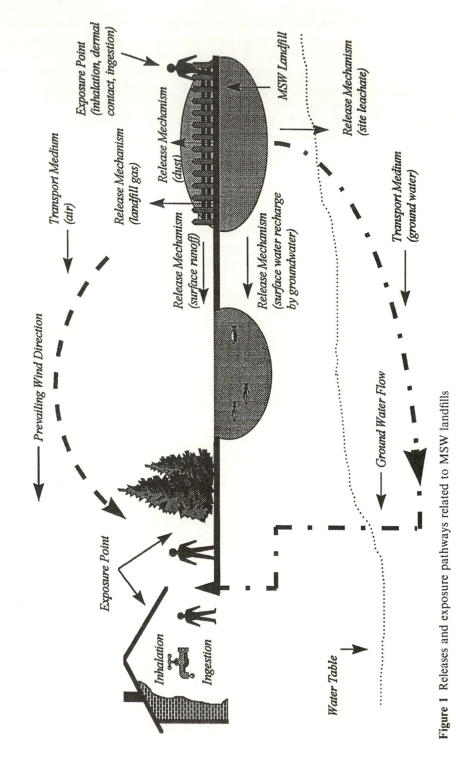

Figure 1 Releases and exposure pathways related to MSW landfills

In terms of their eventual impact on human receptors, the pathways can be summarized as follows:

- *Air inhalation*: This includes inhalation of contaminants as a result of emissions of dust, landfill gas and flare emissions
- *Dermal contact*: This is assumed to occur with on-site landfill operators as a result of exposure to airborne dust. Material adhering to the hands is additionally assumed to be ingested by the operator
- *Soil/dust ingestion*: Ingestion of dust or soil could occur with a landfill operator as a result of contamination of hands. In addition, a portion of the emissions from a landfill will deposit on the soil surface in the vicinity of the site. Off-site receptors could inadvertently ingest soil while playing or working outdoors. Soil is also ingested by grazing animals and forms part of their diet along with pasture
- *Water ingestion*: Chemicals in leachate can enter groundwater and surface water bodies. This water can be directly ingested as potable water, it can support fish consumed by humans or it can be used as irrigation water for market gardens, *etc.*
- *Ingestion of local fish*: The consumption of fish from local streams and rivers is a potentially significant source of indirect exposure to releases from a landfill

Depending on the nature of the activity on or in the vicinity of the landfill site, receptors can be exposed *via* one or a combination of these pathways.

Two additional exposure pathways were considered: the utilization of water containing leachate for irrigation or stock watering, followed by uptake of chemicals present in the leachate into plant or animal products which are eventually consumed by humans; and exposure to releases of micro-organisms from landfill sites. However, on the basis of studies on the effects of leachate on plants, measurement of bacterial and fungal concentrations in air within the boundaries of and in the vicinity of landfill sites,[11-13] and the fact that leachate is not used for irrigation or for stock watering in the UK, these pathways were screened out of the risk assessment.

Receptor Types

On-site receptors are differentiated from off-site receptors. On-site receptors consist of site operators who may be exposed by direct (dermal) contact with the waste and its constituents, dermal contact and inhalation of dust and inhalation of landfill gas. Off-site receptors are assumed to be exposed to dust and associated chemicals released from the landfill, to landfill gas migrating from the site, to emissions resulting from the combustion of landfill gas in flares or gas engines and to leachate *via* ingestion of groundwater. Identification of potentially exposed populations is based on the surrounding land use and documented

[11] D. J. Lisk, *Sci. Total Environ.*, 1991, **100**, 415.
[12] J. Pastor, M. Alia, A. J. Hernandez, M. J. Adarve, A. Urcelay and F. A. Anton, *Sci. Total Environ.*, 1993, suppl. 127.
[13] P. M. Fedorak and R. E. Rogers, *Waste Manage. Res.*, 1991, **9**, 537.

G. Eduljee

sources of demographic information. Patterns of human activity associated with various land uses determine the exposure pathway, duration of exposure and the dose received by the receptor. The distance of the exposed person from the site will also be an important factor in determining the duration and magnitude of the received dose.

Estimation of Dose at the Point of Exposure

Exposure to releases from household waste landfill sites is a function of the estimated concentrations of the released chemicals in the environmental media with which individuals may come into contact (*i.e.* exposure point concentrations) and the duration of contact. The duration of exposure to affected media is estimated for sub-groups within the population, based on certain activity patterns. These activity patterns can be represented by so-called exposure factors which can be expressed as discrete values or as distributions. The exposure equation combines the exposure factors and exposure point concentrations to estimate the received dose. These dose equations are dependent on the route of exposure (*i.e.* ingestion, inhalation and dermal contact). Intakes *via* inhalation and ingestion can be expressed by a general algorithm, as follows:

$$\text{Pathway-specific intake} = \frac{C \times IR \times ET \times EF \times ED \times ABS \times FI}{BW \times AT} \tag{1}$$

where C = concentration of a chemical in the medium of interest; IR = ingestion or inhalation rate ($m^3 h^{-1}$, $mg d^{-1}$, *etc.*); FI = fraction ingested/inhaled from contaminated source; ET = exposure time ($h d^{-1}$); EF = exposure frequency ($d yr^{-1}$); ED = exposure duration (yr); BW = body weight (kg); ABS = fraction of substance absorbed into the body (unitless); AT = time over which the dose is averaged. For off-site exposure, AT is equivalent to the exposure duration (ED) in the case of chemicals with a threshold health effect, and to a lifetime for chemicals with a non-threshold health effect.

For dermal contact, the term IR is replaced by the product of the surface area of skin exposed to the medium (dust or water) and the adherence factor (the amount of the medium that remains on the skin after contact). Intake rates are expressed as picogrammes per kilogram of body weight per day ($pg kg^{-1} d^{-1}$). The algorithms for each exposure pathway are detailed elsewhere.[3]

5 Toxicity Assessment

This stage of the risk assessment process defines the toxicity (*i.e.* the dose–response relationship) for each of the indicator chemicals. For the purposes of quantifying human health risks, chemicals are typically characterized as carcinogens and non-carcinogens. The latter have threshold doses below which they fail to elicit any discernible adverse health effect. Applying safety factors to account for inter-species variability and sensitivity among different members of the population, this threshold dose is converted to an Acceptable Daily Intake

122

Table 1 Summary of toxicological criteria

Release	Pathway	Receptor	Toxicological criteria*
Landfill gas	Inhalation	(1) Operators	(1) Workplace air quality standards
		(2) Public	(2) Statutory air quality standards, or derived long- and short-term air quality standards
Flare/ combustion gas	Inhalation	(1) Operators	(1) Workplace air quality standards
		(2) Public	(2) Statutory air quality standards, or derived long- and short-term air quality standards
Dust	Inhalation	(1) Operators	(1) Workplace air quality standards
		(2) Public	(2) Statutory air quality standards, or derived long- and short-term air quality standards
Dust	Ingestion/ Dermal contact	(1) Operators	(1) Reference dose/ADI/TDI
		(2) Public	(2) Reference dose/ADI/TDI
Leachate/ surface seeps	Ingestion	(1) Operators	(1) Not applicable
		(2) Public	(2) Statutory water quality standards
Landfill gas/ dust	Ingestion of foods	(1) Operators	(1) Not applicable
		(2) Public	(2) Reference dose/ADI/TDI
Landfill gas/ dust	Multiple	(1) Operators	(1) Reference dose/ADI/TDI
		(2) Public	(2) Reference dose/ADI/TDI

*ADI = Acceptable Daily Intake; TDI = Tolerable Daily Intake.

(ADI) or Tolerable Daily Intake (TDI). Carcinogens can be grouped into two categories:

- *Genotoxic*: carcinogens that initiate cancer through an initial effect on DNA
- *Non-genotoxic*: carcinogens that operate through chronic cell damage

Non-genotoxic carcinogens are generally treated as non-carcinogens in so far as a threshold dose is assumed to exist, below which there is no adverse health effect. The method of assessment for genotoxic carcinogens is less clearcut. In the US, it is assumed that a threshold dose does not exist, and therefore any positive dose has the potential to elicit an adverse health effect. The incremental or additional cancer risk to the individual exposed site releases is expressed as an estimated annual or lifetime risk of the chance of mortality from cancer (for

123

example, 1 in 1 million). In the UK the present custom is to treat genotoxic carcinogens in the same manner as non-genotoxic carcinogens; the safety factor approach is applied to both cases.[14] This was the approach taken in the present risk assessment.

The toxicological criteria used in the risk assessment are listed in Table 1. Detailed criteria values and the underlying assumptions are presented elsewhere.[3]

6 Risk Estimation, Evaluation and Control

The steps in risk evaluation are as follows:

- From the exposure assessment stage, the total intake of each chemical into the body is estimated
- The intake of each chemical is compared against an appropriate TDI or reference dose, to obtain a Hazard Index (HI)

Thus, the criterion for 'acceptable' air or water quality in relation to a pollutant released from the landfill is the ratio of the modelled exposure concentration for that pollutant to its air or water quality standard (*i.e.* the HI). The ratio should not exceed 1.0, since this would indicate that the exposure concentration is equal to the standard for air or water quality. The lower the ratio, the less impact the emission has on air or water quality.

There is no agreed method for the assessment of health risks from exposure to mixtures of chemicals. The criterion for acceptability is that the summed Hazard Index should not exceed 1.0, which demonstrates that the safety standard is not exceeded for a mixture of chemicals. Thus:

$$HI_{total} = HI_a + HI_b + HI_c + \ldots < 1.0 \qquad (2)$$

It should be noted that a Hazard Index greater than 1.0 either for single chemicals or the totality of chemicals does not indicate that an adverse health effect will necessarily arise. The ADI or TDI for a chemical itself incorporates safety factors, typically of the order of 10- to 100-fold below the dose representing the so-called 'No Observed Effect Level' (NOEL). Therefore a small exceedance of the criterion indicates that the safety margin is being eroded, and that it would be prudent to manage the exposure pathway which contributes the greatest to the Hazard Index in order to lower the Hazard Index and hence reinstate the margin of safety.

Having assessed and evaluated the risk posed by the landfill site, the final stage of the risk management process is to determine whether the risks require further control and, if so, what these control measures might be. The power of the risk management framework lies in the fact that risk *assessment* and risk *control* are interactive. This is recognized as a key concept in deriving design performance standards and remediation measures for landfills.[15] Thus, if the health risk posed

---

[14] A. McDonald, in *Environmental Impact of Chemicals: Assessment and Control*, ed. M. D. Quint, D. Taylor and R. Purchase, Royal Society of Chemistry, Cambridge, 1996.

[15] Department of the Environment, *Waste Management Paper 26B: Landfill Design, Construction and Operational Practice (Consultation Draft)*, Department of the Environment, London, 1995.

to an off-site receptor by a particular exposure pathway is deemed to merit further reduction, then appropriate operational or engineering measures can be formulated which either break the source → pathway → receptor chain or modify elements of the chain to, for example, reduce emissions, prevent migration or prevent contact with the contaminated medium. Each of these measures is accompanied by a cost, and it is therefore also possible to judge the benefit of the risk control measures in terms of reductions in risk against the cost associated with implementation of these measures.

7 Case Study

Background

The illustrative case study involves an operational household waste landfill site that is shortly to close. The site has a total area of 60 ha and has received 4 million tonnes of household waste. The site has an engineered clay liner of 1 m thickness. Leachate is collected *via* a drainage blanket and a piped drainage system. Landfill gas is collected with 70% efficiency and combusted in a gas engine to generate power. A village is situated to the immediate south of the landfill, and on the southern boundary a farmhouse represents the nearest residence, 50 metres from the boundary of the site. A small industrial estate is located 1 km to the northern boundary of the site. To the north of the site flows a brook into which surface water from the landfill is discharged. The landfill is surrounded by agricultural land, but no irrigation takes place with the surface water from the brook.

The following release pathways were assessed:

- *Releases during operation*: Dust released due to wind erosion, site traffic and placement of waste. Landfill gas migrating from the site and combusted in a gas engine. Transport of leachate to groundwater, and of surface runoff to the brook. Receptors included on-site personnel and off-site public
- *Releases after closure*: Dust released due to wind erosion. Landfill gas migrating from the site and combusted in a gas engine. Transport of leachate to groundwater and of surface runoff to the brook. Receptors included off-site public

The key site and release parameters are discussed below. Detailed worksheets are presented elsewhere.[3]

Release of Dust

The site is designed for the receipt of 4 000 000 tonnes of waste over 15 years. Assuming a 6-day/week, 52 weeks/year operation, this equates to a daily placement of 860 tonnes. This is assumed to generate 100 vehicle movements to the site per day, equating to an average individual payload of 8.6 tonnes. The average vehicle weight is taken as 15 tonnes, the average number of wheels per vehicle as 8, the average speed on the landfill as $20 \, \text{km h}^{-1}$. Assuming that on average the waste vehicle will travel at least to the centre of the site and then return to the gatehouse, the total distance travelled per vehicle on the landfill will be 800 metres, or 80 km for 100 vehicles over 8 hours of operation per day. To this we add a nominal 10 km (total of 110 km) to allow for the operation of a bowser. The emission factor is given in units of kg of dust released per km of distance travelled. The silt content of the cover material is taken as 10%, the moisture content of the waste as 40%, the mean wind speed as $5 \, m \, s^{-1}$, and the number of days per year with at least 0.25 mm precipitation as 100 days.

For dust released during offloading operations and spreading of waste and cover material, the average load is assumed to be 8600 kg, and the total waste offloaded per day as 860 000 kg. The emission factor is given in units of g of dust emitted per kg of waste unloaded.

For dust released as a result of wind erosion, an erosion area of 300 000 m² (50% of the total site area) is assumed. For wind erosion after closure, it is assumed that the site will be progressively capped and restored during the period of operation. It is assumed that on average 5% of the total surface area will be affected by wind erosion, the remaining area being grassed. The affected site area is therefore 30 000 m². For simplicity we assume that the dust contains contamination to the same extent as that for an operational site.

Leachate

Release to Groundwater. The site is designed with a compacted clay liner with a permeability of $10^{-9} \, m \, s^{-1}$. The design allows for a drainage blanket of permeability $10^{-4} \, m \, s^{-1}$ and a piped drainage system with drain spacing of 10 m and a drainage angle of 0.6°. The average annual precipitation rate is taken as $0.5 \, m \, yr^{-1}$ and the runoff coefficient is assumed to be 0.3. We assume that the leachate collection system intercepts 40% of the leachate generated at the site prior to its dispersal through the clay liner. For the closed site, we assume that infiltration will be reduced due to the presence of a cap of permeability $10^{-9} \, m \, s^{-1}$. We therefore allow for a runoff coefficient of 0.7 during the period of closure. We assume that 40% of the leachate penetrating the cap will be intercepted by the leachate collection system.

In order to calculate leachate flow to groundwater using the LANDSIM model[10] the clay liner is treated as an initial unsaturated zone of 1 m thickness, a dispersivity coefficient of 0.1 m, a dry bulk density of $2 \, g \, cm^{-3}$, a water content of 0.3 and an organic carbon fraction of 0.01. The unsaturated zone extends 8 m beneath the base of the clay liner, and has a dispersivity coefficient of 1 m, a water content of 0.3, an organic carbon fraction of 0.02 and a dry bulk density of $2 \, g \, cm^{-3}$. The saturated zone has a dispersivity coefficient of 10 m, a porosity of 0.3, a hydraulic gradient to the off-site well of 0.05 and a hydraulic conductivity of $10^{-6} \, m \, d^{-1}$. Off-site receptors are assumed to be present at the farmhouse (50 m from the site boundary) and at the village (1 km from the site boundary).

Seeps to Surface Water. The landfill design includes perimeter drains to collect any seeps and surface water leaving the site. As a worst case it is assumed that these flows are equated to the release of 10% of the leachate produced by the site,

Table 2 Exposure assumptions for calculation of on-site inhalation exposure*

Receptor	IR/ $m^3 h^{-1}$	ET/h d^{-1} Landfill	Perimeter	d yr^{-1}	ED/ yr	AT/ yr	ABS and FI	BW/ kg
Delivery drivers	1	1.5	0	250	25	25	1	70
On-site drivers	1	7	1	250	25	25	1	70
Site labourers	1	7	1	250	25	25	1	70
Gas plant engineers	1	1	7	250	25	25	1	70
Monitoring technicians	1	3	0	150	25	25	1	70
Supervisors	1	2	6	250	25	25	1	70
Office staff	1	0	8	250	25	25	1	70
Contract staff	1	6	2	250	25	25	1	70

*See Table 3 for explanation of column headings.

in unattenuated form. The flow rate of the receiving brook is assumed to be $80\,000\,m^3\,d^{-1}$.

Landfill Gas

Migrating Gas. At an assumed landfill gas generation rate of $5\,m^3\,tonne^{-1}\,yr^{-1}$ and a daily placement of 860 tonnes of waste, $260\,000\,m^3$ of landfill gas will be freshly generated each year. Since the site is shortly to close, we consider a scenario in which the site will contain $4\,000\,000$ tonnes of household waste, generating 20 million m^3 of gas annually for a further period of 15–20 years (*i.e.* for a total period from commencement of operations of 30 years). Assuming that 70% of the landfill gas will be extracted and combusted, 6 million $m^3\,yr^{-1}$ of gas will migrate to atmosphere, equating to a flux of $20\,m^3\,m^{-2}\,yr^{-1}$ over the entire site.

Combusted Gas. Of the total landfill gas generated at the site, we assume 70% will be combusted in a gas engine. Assuming continuous operation of the gas engine, this equates to a flow rate of landfill gas to the gas engine of $0.44\,m^3\,s^{-1}$. The physical stack height is assumed to be 10 m. We assume unstable conditions and a wind speed of $5\,m\,s^{-1}$. Assuming $10\,m^3$ of flue gas per m^3 of landfill gas combusted, the gas flow rate exiting the stack will be $4.4\,m^3\,s^{-1}$ at a velocity of $15\,m\,s^{-1}$.

Exposure Assessment

Inhalation. For on-site exposure, two inhalation scenarios are derived to encompass the types of personnel typically present on a landfill site:

(a) Workers on or close to the operational face of the landfill, assumed to be the most exposed to dust (banksmen, delivery drivers and on-site drivers). The box model described in Section 3 was used, the width of the box being 15 m.

Table 3 Exposure assumptions for calculation of off-site inhalation exposure

Parameters	Adult resident	Child resident	School child	Infant resident	Off-site worker	Adult farmer
IR: inhalation rate ($m^3 h^{-1}$)*	0.62/ 0.70	0.40/ 0.45	0.64/ 0.74	0.23/ 0.26	0.62/ 0.70	0.62/ 0.70
ET = exposure time ($h d^{-1}$)*	21/3	20/4	6/2	23/1	4/4	14/10
EF = exposure frequency ($d yr^{-1}$)	330	330	205	330	235	330
ED = exposure duration (yr)	30	5	10	0.5	20	40
BW = body weight (kg)	70.1	15	42.5	8.5	70.1	70.1
LT = lifetime (d)	27 375	27 375	27 375	27 375	27 375	27 375
AT = averaging time (d)	10 950	1825	3650	183	7300	14 600

*Indoor/outdoor.

(b) Workers elsewhere on the landfill, including the site perimeter and the site office (all other site personnel). The box model was used with an increased box width of 100 m to allow for linear sources (roads) and other general sources that are more dispersed than in the offloading area.

The values inserted into eqn. (1) for on site are listed in Table 2. For off-site exposure, the values for the parameters in eqn. (1) are listed in Table 3.

We differentiate between outdoor and indoor exposure by applying a factor of 0.7 to the outdoor ambient air concentration of a chemical to allow for lack of penetration of dust into the home. The same principle applies to off-site workers, where 4 hours of the 8 hour shift are assumed to be spent out of doors.

Dermal Contact. Dust concentrations are obtained from PM_{10} dust concentrations calculated for the purpose of assessing inhalation exposure. While this approach will underestimate the degree of dermal contact with dust for on-site receptors (because dust particles of larger size will also be disturbed during landfill activities), this is to some extent compensated for by the conservative exposure assumptions. For off-site receptors this approach should not underestimate dermal contact to a significant extent because larger particles would be subject to settlement in the immediate vicinity of the landfill site. The product of skin surface area and adherence factor is set at 742 mg for all activities performed 'on landfill' (*i.e.* all site personnel except for office staff) and at 605 mg for all activities performed 'on perimeter' (*i.e.* all site personnel except for delivery drivers and monitoring technicians). For the parameter ABS, a value of 0.01 was adopted for all metals and a value of 0.1 was adopted for all organics.

Ingestion of Dust/Soil and Water. 'On landfill' workers are assumed to ingest dust/soil as for adult farmers (74 mg kg^{-1}), while office staff and other 'on

Table 4 Case study: Total Hazard Index for on-site receptors (operational/closed)*

Receptor type	Inhalation	Dermal contact	Dust ingestion	Total Hazard Index
Delivery driver	0.19/(0.002)	0.03/(0.028)	0.003/(0.003)	0.22/(0.033)
On-site driver	0.88/(0.01)	0.15/(0.15)	0.014/(0.014)	1.03/(0.17)
Site labourer	0.88/(0.01)	0.15/(0.15)	0.014/(0.014)	1.03/(0.18)
Gas plant engineers	0.14/0.01	0.13/0.13	0.008/0.008	0.27/0.15
Monitoring technicians	0.23/0.002	0.034/0.034	0.004/0.004	0.26/0.04
Supervisors	0.26/(0.01)	0.13/(0.13)	0.009/(0.009)	0.40/(0.15)
Office staff	0.02/(0.01)	0.12/(0.12)	0.007/(0.007)	0.15/(0.14)
Contract staff	0.75/(0.01)	0.14/(0.14)	0.013/(0.013)	0.91/(0.17)

*The hypothetical presence of staff on site after landfill closure is represented by bracketed values for the Hazard Indices.

perimeter' personnel are assumed to ingest dust/soil as for adult residents (37 mg kg^{-1}). Off-site adult receptors are assumed to ingest 2 litres of water per day. Adolescent, child and infant receptors are assumed to consume 2.0, 0.8 and 0.7 litres of water a day, respectively.[16]

Results and Discussion

The Hazard Indices representing the impact of cumulative doses of the indicator chemicals are presented in Table 4. After closure of the landfill, all operatives are assumed to have remained on site in order to enable a comparison to be made between open and closed sites—the hypothetical presence of staff on site is represented by the bracketed values in Table 4.

During operation, the inhalation pathway represents the exposure route with the greatest potential for chemical uptake, with lead in dust accounting for about 80% of the Hazard Index for inhalation, and about 75% of the total Hazard Index. While the latter is only marginally elevated above the criterion for acceptability, dust suppression measures such as the wearing of face masks during particularly dusty periods and dousing of waste and site roads with water are nevertheless prudent precautionary measures to be considered for the protection of on-site drivers, site labourers and contract staff.

The closure of the landfill results in a significant fall in the Hazard Index for on-site inhalation such that this pathway now becomes a minor exposure route. Dermal contact is the main exposure route for site operatives after landfill closure; this exposure pathway is readily controlled through the use of personal protection such as overalls.

The Hazard Indices relating to off-site receptors are summarized in Table 5. As in the case of on-site receptors, the operational and closed scenarios are compared in the same table. The results indicate that the exposure experienced by all off-site receptors is well within the criterion of acceptability. Within the total

[16] US EPA, *Exposure Factors Handbook*, EPA/600/8-89/043, Office of Health and Environmental Assessment, Washington, DC, 1989.

Table 5 Case study: Total Hazard Index for off-site receptors (operational/closed)

Receptor type	Inhalation	Water ingestion	Fish ingestion	Total Hazard Index
Adult resident	0.0003/0.0002	0.02/0.02	0.01/0.005	0.03/0.025
Child resident	0.0010/0.0005	0.06/0.06	0.01/0.005	0.07/0.065
School child	0.0001/<0.0001	0.01/0.01	0.01/0.005	0.02/0.015
Infant	0.0010/0.0005	0.04/0.04	0.02/0.01	0.06/0.05
Off-site worker	<0.0001/<0.0001	0.02/0.02	0.01/0.005	0.03/0.025
Adult farmer	0.0004/0.0002	0.02/0.02	0.01/0.005	0.03/0.025

Hazard Index, the dominance of the groundwater/ingestion pathway in driving potential risks from landfills is evident. Reductions in chemical uptake *via* inhalation exposure and fish ingestion, while significant in themselves relative to the operational scenario, are insufficient to effect large reductions in the total Hazard Index after site closure.

The groundwater ingestion route represents an extreme exposure. The travel time for the leachate (*i.e.* chloride ion) through the unsaturated zone is approximately 30 years, while the travel time for metals such as lead is over 1000 years. The travel time for lead to the off-site receptor wells is also in the order of 1000 years. Therefore the groundwater pathway does not in fact represent a potential threat to human health over timescales of concern.

Uncertainty in the Total Hazard Index

An uncertainty analysis was conducted with the proprietary @RISK software package, which utilizes Monte Carlo simulation to describe uncertainty in the form of probability distribution functions. The results of the Monte Carlo simulation indicated that a factor of approximately four separated the 5th percentile value and the 95th percentile value of the total Hazard Index; the 95th percentile and the mean value of the total Hazard Index differed by a factor of two.

The 95th percentile level represents a conservative exposure scenario, the cumulative effect of input parameters at the high end of their range. If a decision maker chooses to regulate the risk experienced by an exposed individual at the 95th percentile, it is of interest to examine how the total Hazard Index at this level varies relative to the criterion for acceptability, of 1.0. Inspection of Table 4 indicates that for delivery drivers and site labourers the mean HI was already above 1.0, giving a 95th percentile total HI in excess of 2.0. The mean HI for contract staff is 0.95, again causing the 95th percentile value to exceed 1.0. As discussed above, protection of on-site staff from dust inhalation would constitute a prudent precautionary measure. For the remaining on-site receptors the criterion for acceptability is not exceeded at the 95th percentile level, indicating acceptable exposure.

The 95th percentile Hazard Index for all off-site receptors is well below the criterion of acceptability of 1.0. Since the 95th percentile represents very conservative exposure levels, the environmental setting and existing controls on the site are sufficiently protective towards public health.

Table 6 Case study: Total Hazard Index (child resident) and cost estimates for risk control measures

Risk control measure	Total Hazard Index	Estimated cost	Incremental cost per unit reduction in HI
Baseline	0.97	—	—
Option 1 (increased leachate draw-off)	0.45	£420 000–£720 000	£916 500
Option 2 (regrading the surface)	0.76	£500 000	£2 381 000
Option 3 (installing a pump and treat system)	0.20	£200 000–£400 000	£389 610

Risk Control

Table 5 indicates that for all the off-site receptors considered the total Hazard Index is well below the threshold of acceptability, implying that existing landfill design, control and management measures provide for adequate protection of public health, and that additional risk control measures are therefore not required. Nevertheless, for illustrative purposes only, we demonstrate the application of risk reduction measures in the context of landfill releases by considering a hypothetical situation in which the objective is to lower the risk of potential harm *via* the principal pathway contributing to the total Hazard Index, namely ingestion of groundwater. For the closed phase of the landfill, we assume this can be achieved by the following means:

Option 1. *Increased leachate draw-off.* Increased draw-off of leachate could be achieved either by retrofitting additional drainage runs within the cells to improve drainage to draw-off points, or alternatively by installing a number of dewatering wells with direct draw-off from each well. Assuming an additional requirement for 30–40 wells and associated pipework, the capital cost would typically be in the range of £400 000–£700 000 with an annualized operating and maintenance cost of £20 000. Since a leachate treatment system is already in place, we assume that the present system can accommodate the increased volumes associated with this control option. However, if a leachate treatment system was required, an additional capital cost of £1 000 000–£3 000 000 is indicated. An alternative is to intercept the leachate at the downgradient edge of the site.

Option 2. *Regrading the surface.* Assuming minor regrading of the surface of the landfill and the installation of an artificial membrane and soil cover at a unit cost of £15 per m², Option 2 would cost in the region of £500 000.

Option 3. *Installing a groundwater pump and treat system.* The cost of a system to extract the groundwater and process to reduce the metal content before it reaches the public supply would be in the region of £100 000–£300 000 in the event that the present leachate treatment system can accommodate the increased flows. If a dedicated treatment system is required, then the additional capital cost would be in the region of £300 000–£500 000. We add a further £100 000 in annualized operational and maintenance costs.

Monitoring of groundwater quality will be an integral part of the overall management of the site. This cost is therefore not included as an additional item to be considered during risk control.

Implementation of Options 1–3 will result in reduced leachate generation volumes relative to the base case. The altered assumptions for Options 1–3 were inserted into the LANDSIM model. The effect of these changes on the total Hazard Index is illustrated in Table 6 for the child resident, discounting depreciation on capital costs.

Option 2 is comparable in cost to Option 1 but does not produce a significant reduction in the total Hazard Index. Option 3 is the least costly and results in a significantly lower hazard index, but does not address leachate production at source. Option 1 addresses leachate production at source and produces a lower total Hazard Index.

An alternative depiction of cost benefit is to examine the incremental cost for a unit decrease in the Hazard Index. This can be achieved by dividing the cost of the option by the difference between the HI of the base case and the HI of the option. The higher the incremental cost, the less is the 'value for money' for the particular option in terms of risk reduction. This calculation is also shown in Table 6. Inspection of the costs suggests that of the options for leachate reduction, Option 3 is the least costly and the most effective in terms of cost per unit reduction in the Hazard Index. However, in the long term the implementation of this option as the sole risk control measure would not be recommended since it does not reduce leachate production at source. The preferred risk control option is therefore Option 1, perhaps in conjunction with a less extensive regrading of the site since the cumulative effect will be a further reduction in leachate generation.

8 Summary

The case study illustrates how a risk assessment procedure can be used to obtain information on which pathways are most significant in terms of the overall exposure experienced by a particular receptor, and within these pathways which chemicals dominate the Hazard Index. Based on this information, targeted risk control measures and associated costs can be identified.

There are several limitations to the study, limitations which are generally common to other processes for which public health risk assessment is undertaken. Chief amongst these are the following:

- The lack of comprehensive site-specific data on indicator chemical concentrations in releases from household waste landfill sites. The data used in the case studies are generic in nature, though the uncertainty analysis does allow for a degree of variability in these and other model inputs.
- The lack of information on the concentrations of chemicals in the ambient environment external to the site. This is particularly true for ambient air quality, but also holds for groundwater and surface water quality. It is therefore not possible to calibrate the model output against measured site-specific environmental concentrations for any of the exposure pathways examined in this risk assessment.

- The transport algorithms developed for landfill gas releases and for PM_{10} particulates[3] have yet to be validated. The LANDSIM model for the simulation of leachate migration is a simplification of the complex mechanism known to exist in field conditions.[10]

Despite these simplifications and limitations of data availability, the risk assessment framework can be a useful aid to decision making, especially when alternative risk control options are assessed for the same site.

9 Acknowledgements

The author is grateful to the Environment Agency for funding this work and granting permission for publication. The views expressed in this chapter are those of the author and do not necessarily represent those of the Environment Agency. The author especially wishes to thank Dr Jan Gronow and Dr Martin Meadows of the Waste Regulation Policy Group for the valuable assistance and advice they have offered throughout the study.

10 Appendix

Table A1 Indicator chemicals in landfill gas, and associated default concentrations in raw gas, flared gas and gas engine emissions[17]

Chemical	Raw gas/ $mg\,m^{-3}$	Flared gas/ $mg\,m^{-3}$	Gas engine emissions/ $mg\,m^{-3}$
Vinyl chloride	8.3	0.5	0.5
Carbon tetrachloride	0.7	0.003	0.003
Chloroform	1.0	0.007	0.007
Methyl chloride	10	0.9	0.9
1,2-Dichloroethane	21	0.9	0.9
1,1-Dichloroethene	5.0	0.9	0.9
Trichloroethene	37	0.9	0.9
Tetrachloroethene	22	9.0	9.0
Benzene	12	0.8	0.8
Ethylbenzene	50	15	15
Dioxins/furans (ng I-TEQ m^{-3})*	0.34	0.022	0.07
PM_{10}*	N/A	10	10

*I-TEQ = International Toxic Equivalents; PM_{10} = Particulate Matter of less than $10\,\mu m$ in diameter.

[17] H. D. Robinson, *A Review of the Composition of Leachates from Domestic Wastes in Landfill Sites*, Report No. DE0918A/FR1, Department of the Environment, London, 1995.

Table A2 Indicator chemicals and their proposed default concentrations in leachate[3,17-19]

Chemical	Fresh wastes*/mg l^{-1}	Aged wastes*/mg l^{-1}
Aluminium	0.1	0.1
Antimony	2.0	2.0
Arsenic	0.008	0.008
Cadmium	0.013	0.013
Chromium	0.14	0.14
Copper	0.04	0.005
Lead	0.4	0.14
Magnesium	252	185
Manganese	27	2.1
Mercury	0.0002	0.0001
Nickel	0.6	0.1
Selenium	0.04	0.04
Silver	0.05	0.05
Zinc	21.5	0.4
Atrazine	4.04×10^{-3}	4.04×10^{-3}
Dichlorvos	1.59×10^{-3}	1.59×10^{-3}
Hexachlorobenzene	0.09×10^{-3}	0.09×10^{-3}
Lindane	3.37×10^{-3}	3.37×10^{-3}
PCB	0.08×10^{-3}	0.08×10^{-3}
Simazine	3.55×10^{-3}	3.55×10^{-3}
1,2,3-Trichlorobenzene	0.58×10^{-3}	0.58×10^{-3}
1,3,5-Trichlorobenzene	0.19×10^{-3}	0.19×10^{-3}
Total organotin	0.53×10^{-3}	0.53×10^{-3}

*Fresh wastes: leachate from an operational site; aged wastes: leachate from a closed site.

[18] H. D. Robinson and J. R. Gronow, in *Proceedings of the Fourth International Landfill Symposium*, University of Cagliari, Sardinia, 1993.

[19] Department of the Environment, *Waste Management Paper 26A: Landfill Completion*, Her Majesty's Stationery Office, London, 1993.

Chemical	Concentration in dust/household waste/ mg kg^{-1}
Aluminium	9000
Antimony	3.4
Arsenic	4.7
Cadmium	0.74
Chromium	370
Copper	120
Lead	295
Magnesium	1800
Manganese	350
Mercury	0.03
Nickel	28
Silver	0.5
Zinc	1300
Dioxins/furans	0.0063
Atrazine	0.1
Dichlorvos	0.1
Hexachlorobenzene	0.125
Lindane	0.1
PCB	0.2
Simazine	0.1
1,2,3-Trichlorobenzene	0.125
1,3,5-Trichlorobenzene	0.125
Total organotin	0.1

Table A3 Indicator chemicals and their proposed default concentrations in dust/household waste[3,20,21]

[20] Department of the Environment, *The Technical Aspects of Controlled Waste Management: The National Household Waste Analysis Project, Phase 2, Volume 3*, Report No. CWM/087/94, Department of the Environment, London, 1994.

[21] B. Johnke and E. Stelzner, *Waste Manage. Res.*, 1992, **10**, 345.

Aquatic Risk Assessment and Risk Management for Pesticides

STEVE MAUND AND NEIL MACKAY

1 Introduction

Ecological risk assessment for agricultural pesticides (insecticides, herbicides, fungicides, rodenticides, plant and insect growth regulators) needs to be comprehensive, because the purpose of these compounds is to control organisms which may have closely related non-target counterparts in the environment. Consequently, the data generated for pesticide registration are substantial and include methods covering all aspects of the ecological effects and environmental fate of the compound. Studies are performed to determine the effects of pesticides on all of the major non-target organism groups, *e.g.* birds, mammals, worms, terrestrial arthropods, terrestrial plants, fish, aquatic invertebrates, aquatic plants and soil micro-organisms. In addition, extensive data on the environmental fate of the compound are produced, *e.g.* degradation in soil, water and sediment, photolysis, hydrolysis, adsorption, residues in the crop. The role of ecological risk assessment is to integrate these effects and fate data into an analysis of the potential for exposure and unacceptable adverse effects on non-target organisms, and particularly to put these into the context of agricultural use.

Approaches to ecological risk assessment for pesticides and the science on which these are based have been developing rapidly over the last decade. In this article, current risk assessment and management techniques for pesticides as they relate to aquatic ecosystems will be reviewed.

Principles of Aquatic Risk Assessment—Hazard, Risk and Tiered Approaches

The terminology and phraseology used in risk assessment is often inconsistent and confusing. The reader is referred to the Organization for Economic Cooperation and Development's (OECD) definitions for a useful synthesis of the terms.[1] In particular, risk and hazard are two terms that are often used inconsistently in ecotoxicology. *Hazard* usually describes the inherent action of

[1] Organization of Economic Cooperation and Development, *Report of the OECD Workshop On Environmental Hazard/Risk Assessment*, Environment Monographs No. 105, OECD Environment Directorate, Paris, 1995.

137

the chemical on an organism (in this case, its toxicity). In aquatic ecotoxicology, hazard is usually categorized by the concentration of chemical which elicits a certain response after a prescribed time interval. Examples of hazard measurements include:

- The median effective or median lethal concentration (EC_{50} or LC_{50}), which are estimates of the concentration which will affect 50% of the test population
- The no observed effect concentration (NOEC), which is the concentration which causes no significant difference compared to an untreated control

Hazard measurements are used to categorize the degree of effect a compound might be expected to exert. For example, a compound which has a 96 h LC_{50} of $10 \, \mu g \, l^{-1}$ to fish may be termed as being a highly toxic (*i.e.* hazardous) compound for fish because it has the potential to be toxic at low concentrations. These measures of hazard are also called *effect concentrations*.

Risk can be simply defined as an estimation of the potential for a certain hazard to be realised (*i.e.* the likelihood that the effect concentration will occur in the environment). In aquatic risk assessment, risk is most commonly quantified by comparing effect concentrations to an estimated concentration of the chemical in a model aquatic ecosystem such as a pond or a ditch—the *exposure concentration*. The proximity of the effect and exposure concentration is then used to estimate the potential for an effect to occur—*risk characterization*. Generally, if the exposure concentration is greater than or equal to the effect, then impacts might be expected to occur. In practice, *safety factors* are often applied, typically so that exposure and effect concentrations should differ by a predetermined factor greater than one—the larger the factor, the more conservative the assessment. The likelihood as to whether impacts would be realized in the environment, *risk estimation*, depends also on the conservatism of the approaches used. For example, if the model aquatic system used to determine exposure has proportions such that dilution is less than the majority of natural ecosystems, then risks may be overestimated.

There are two principal approaches to risk assessment:

- *Determininistic* approaches, which have a standard exposure and effect scenario to develop single effect and exposure concentrations which are then used to develop a *risk quotient*. This is the most commonly used for pesticides and will be discussed in more detail below
- *Probabilistic* approaches, which develop probability distributions of exposure and effect concentrations and then either determine the degree of overlap between the distributions or the probability of a certain effect concentration being exceeded. This has only recently begun to be considered for agrochemical risk assessment, and the application of these techniques is still under discussion. Probabilistic approaches will not be considered in detail here (more information on probabilistic approaches can be found elsewhere[2-5]).

[2] Health Council of the Netherlands, *Network*, 1993, **6/7**, 8.

[3] Society of Environmental Toxicology and Chemistry, *Pesticide Risk and Mitigation: Final Report of the Aquatic Risk Assessment and Mitigation Dialog Group*, Society of Environmental Toxicology and Chemistry Foundation for Environmental Education, Pensacola, FL, 1994.

Pesticide risk assessment schemes typically have a number of steps of increasing sophistication and environmental realism. This is known as the *tiered approach*. The premise of a tiered approach is that early tiers should be conservative, utilizing simple laboratory data generated under worst-case conditions with estimates of exposure generated from 'worst-case' exposure models. If these preliminary risk characterizations indicate a potential for unacceptable effects, further evaluation at the next tier is required. The higher tiers typically include more realistic estimates of effects and exposure, ultimately with assessment of effects under realistic use conditions, environmental monitoring or a landscape-level of exposure assessment. Generally, higher tier evaluations involve increasing levels of scientific sophistication and the generation of considerably more data. The process is one of iteration, where both effects and exposure estimates are refined together. Although approaches to effect and exposure characterization are described separately below, in practice they are carried out side-by-side.

The purpose of the tiered approach is to act as a positive *trigger*. If a compound is designated safe by extreme worst-case estimates of exposure and effects in a preliminary risk assessment, then it can be concluded with a good deal of certainty that, under typical conditions, there will be no adverse effects. If the compound fails a preliminary assessment, this triggers further evaluation at a higher level, with appropriate refinement of the risk assessment. It is only after evaluation at the highest tier (*i.e.* under realistic environmental conditions) that it can be firmly concluded that a compound may present a significant risk to the environment. However, indications of low risk can be concluded much earlier in the evaluation because the scheme is conservative. This tiered approach is therefore of benefit to both pesticide developer and regulatory authorities (and hence the public) because it allows the early identification of low-risk compounds, enabling appropriate effort to be focused on issues which might raise environmental concerns. Not all pesticides therefore will require higher tier or refined risk assessment under all circumstances.

The general principles involved in risk assessment are illustrated in Figure 1. This indicates that risks should be assessed by examining the relationship between the inherent toxicity of a compound to an organism and its likely exposure in the environment. Potential risks should be evaluated in relation to the likely ecological consequences of any such impact. For example, minor impacts on an organism with a short generation time and high reproductive output may be of little significance to the ecosystem as a whole, whereas impacts on a top predator may be very significant. Risk assessment should also be tempered by considering the *benefit* associated with the chemical (*i.e.* impacts on yields if not used, impacts of alternatives, public health uses). Although often only

[4] K. R. Solomon, D. B. Baker, P. Richards, K. R. Dixon, S. J. Klaine, T. W. La Point, R. J. Kendall, J. M. Giddings, J. P. Giesy, L. W. Hall Jr., C. P. Weisskopf and M. Williams, *Environ. Toxicol. Chem.*, 1996, **15**, 31.

[5] D. Laskowski, A. S. McGibbon, P. L. Havens and S. A. Cryer, in *Pesticide Movement to Water*, ed. A. Walker, R. Allen, S. W. Bailey, A. M. Bailey, A. M. Blair, C. D. Brown, P. Guenther, C. R. Leake and P. H. Nicholls, Monograph No. 62, British Crop Protection Council, Farnham, 1995, p. 137.

Figure 1 Principal processes of risk–benefit analysis for pesticides

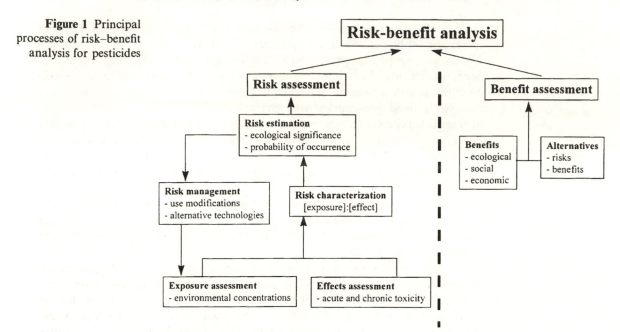

treated as a minor or incidental consideration, benefit analyses are equally an important component of pesticide regulation as risk assessment.

Standardized tier-testing procedures to measure effects and estimate exposure of pesticides in aquatic ecosystems have been developed over recent years and are still going through a process of refinement. Recent years have seen the publication of new procedures for assessing pesticides in both the USA[3,6,7] and Europe.[8] Although there are differences in the details of these methods, in approach they are similar and include the principles described in Figure 1. In Japan there is also a tiered approach to toxicity testing, but at present there is less emphasis on determining exposure calculations—the assessment process for pesticides is principally based on hazard. The remainder of this article will describe some of the approaches to developing effects and exposure data in the USA, the EU and Japan, and their potential uses in pesticide risk assessment.

2   Effects Assessment

The purpose of effects assessment is to describe the concentrations at which aquatic organisms are likely to be affected. Initial laboratory studies investigate endpoints such as mortality or immobilization (expressed as median lethal or

---

[6] D. J. Urban and N. J. Cook, *Ecological Risk Assessment. EPA Hazard Evaluation Division Standard Evaluation Procedure*, United States Environmental Protection Agency Document EPA-540/9-85-001, Washington, DC, 1986.

[7] World Wildlife Fund, *Improving Risk Assessment under FIFRA. Report of the Aquatic Effects Dialogue Group*, World Wildlife Fund, Washington, DC, 1992.

[8] Society of Environmental Toxicology and Chemistry—Europe, *Procedures for Assessing the Environmental Fate and Ecotoxicity of Pesticides*, ed. M. Lynch, SETAC-Europe, Brussels, 1995.

median effective concentrations—LC/EC_{50}) over short exposure periods (acute toxicity tests). Further studies evaluate the impacts over longer time periods on other aspects of the organism's life-history (chronic studies). Endpoints typically evaluated in chronic studies include growth and reproduction. Laboratory studies allow inferences about the likely impacts of pesticides on individuals and, to a limited extent, populations of species. To evaluate potential impacts on communities or ecosystems, field studies (mesocosms and microcosms) which contain multiple species in a semi-natural setting are typically used to evaluate effects. A range of triggers which determine whether particular studies are required, depending on the inherent properties or the use pattern of the compound, are usually set out in pesticide regulation schemes. The triggering of studies used in effects assessments in various parts of the world are discussed below.

Core Data Requirement for Active Ingredients

The core data requirements for the assessment of pesticide active ingredients are acute toxicity tests on *Daphnia* and fish (measuring immobilization and mortality, respectively), and evaluation of effects on algal growth (more information on international requirements is available[9]). Details of the requirements for the EU, USA and Japan are shown in Table 1. The EU and USA require testing of both cold-water (typically rainbow trout, *Oncorhynchus mykiss*) and warm-water fish (a variety of warm-water species are acceptable).

In practice, studies conducted to the various guidelines are broadly interchangeable for registration in the USA and EU (see Cooney[10] for a comprehensive review of toxicity methodologies), and much effort in recent years has gone into developing internationally harmonized procedures for the conduct of these tests. The OECD has been at the forefront of these activities.[11] At present, Japan has its own recommended methods for conducting these studies, and there are slight differences in the organisms used, and the conduct and reporting of these studies.[12]

Further Study Requirements on Active Ingredients Triggered by Pesticide Properties or Uses

In addition to these basic data requirements, further studies on aquatic organisms may be triggered, depending on the properties and use pattern of the compound (Table 2). These studies typically address concerns over potential longer-term exposure, by addressing chronic endpoints such as growth and reproduction. The triggering for these studies varies between regulatory authorities,

[9] Organization for Economic Cooperation and Development, *Data Requirements for Pesticide Registration in OECD Member Countries: Survey Results*, OECD Series on Pesticides No. 1, Paris, 1994.

[10] J. D. Cooney, in *Fundamentals of Aquatic Toxicology: Effects, Environmental Fate and Risk Assessment*, ed. G. M. Rand, Taylor and Francis, Washington, DC, 2nd edn., 1995, p. 71.

[11] Organization for Economic Cooperation and Development, *Guidelines for Testing of Chemicals*, OECD, Paris, 1987.

[12] Y. Nishiuchi, *Jpn. Pestic. Inf.*, 1974, **19**, 15.

Table 1 Core data requirements for aquatic effects assessments for pesticides

Test species	Test endpoint	Endpoint used in evaluations*		
		EU	USA	Japan
Green alga (*e.g. Selenastrum capricornutum*)	Growth rate and biomass	72 h EC$_{50}$s	120 h EC$_{50}$s	nr
Water flea (*Daphnia magna*; other species also acceptable for Japan)	Immobilization	48 h EC$_{50}$	48 h EC$_{50}$	3 h LC$_{50}$ (with adult)
Cold water fish (*e.g.* Rainbow trout [*Oncorhynchus mykiss*])	Mortality	96 h LC$_{50}$	96 h LC$_{50}$	nr
Warm-water fish (*e.g.* bluegill sunfish [*Lepomis macrochirus*], fathead minnow [*Pimephales promelas*], carp [*Cyprinus carpio*])	Mortality	96 h LC$_{50}$	96 h LC$_{50}$	nr
Japanese carp (*Cyprinus carpio*)	Mortality	nr	nr	48 h LC$_{50}$

*nr = not required; LC/EC$_{50}$ = median lethal/median effective concentration.

Table 2 Studies triggered on the basis of preliminary toxicity or *via* use/type of the compound

Test species	Test endpoint	Endpoint used in evaluations*		
		EU	USA	Japan
Blue-green alga *Anabaena flos-aquae*	Growth and reproduction	72 h EC$_{50}$s and NOECs	120 h EC$_{50}$s	nr
Freshwater diatom *Navicula peliculosa*	Growth and reproduction	nr	120 h EC$_{50}$s	nr
Marine diatom *Skeletonema costatum*	Growth and reproduction	nr	120 h EC$_{50}$s	nr
Floating macrophyte *Lemna gibba*	Growth	nr	14 d EC$_{50}$s	nr
Water flea *Daphnia magna*	Survival and reproduction	21 d NOEC	21 d MATC	21 d NOEC
Sheepshead minnow *Cyprinodon variegatus*	Mortality	nr	96 h LC$_{50}$	nr

Table 2 (*cont.*)

Test species	Test endpoint	Endpoint used in evaluations*		
		EU	USA	Japan
Mysid shrimp *Mysidopsis bahia*	Mortality	nr	96 h LC$_{50}$	nr
Pacific oyster (embryo larvae) *Crassotrea gigas*	Larval deformity	nr	48 h EC$_{50}$	nr
Aquatic macrophytes	Not specified	No current guideline	nr	nr
Fathead minnow *Pimephales promelas*	ELS survival and growth	32 d NOEC	32 d MATC	nr
Sheepshead minnow *Cyprinodon variegatus*	ELS survival and growth	nr	32 d MATC	nr
Mysid shrimp *Mysidopsis bahia*	Survival and reproduction	nr	28 d MATC	nr
Fish bioconcentration *Lepomis macrochirus*	Case by case	BCF	BCF	BCF
Aquatic insect (species not specified)	Chronic (not specified)	No current guideline	nr	nr
Gastropod mollusc (species not specified)	Chronic (not specified)	No current guideline	nr	nr
Rainbow trout *Oncorhynchus mykiss*	Mortality	nr	nr	48 h LC$_{50}$
Rice fish *Oryzias latipes*	Mortality	nr	nr	48 h LC$_{50}$
Asian loach *Misgurnus anguillicaudatus*	Mortality	nr	nr	48 h LC$_{50}$
Prawn acute *Neocaridinia dendiculate*	Mortality	nr	nr	48 h LC$_{50}$
Shellfish *Corbicula meana*	Mortality	nr	nr	48 h LC$_{50}$
Fish deformity studies (rice fish, *Oryzias latipes*)	Larval malformation	nr	nr	NOEC

*nr = not required; LC/EC$_{50}$ = median lethal/median effective concentration; NOEC = no observed effect concentration; MATC = maximum acceptable toxicant concentration; ELS = early-life stage; BCF = bioconcentration factor.

but is often based on either the inherent toxicity of the compound or the potential for longer-term exposure in aquatic environments.

For pesticide registration in the EU the requirements for higher tier tests are described in the technical annexes of Directive 91/414/EEC. There is still a great deal of on-going discussion about the interpretation of these requirements, which in some cases are fairly vague.

Studies of an additional taxonomic group of algae are required for herbicides. Appropriate tests can be conducted with either a freshwater diatom (*e.g. Navicula*) or blue-green algae (*e.g. Anabaena*) for which standard methodologies are available. A test on aquatic higher plants (macrophytes) is also required for herbicides, and studies on the duckweed *Lemna* (for which there is an US EPA method and an OECD method in development) can fulfil this requirement.

For aquatic invertebrates, a chronic study on *Daphnia* is required unless it can be justified that repeated or continued exposure of the compound to aquatic organisms will not occur. The repeated exposure trigger is met for all pesticides which are used more than once, and therefore this requirement will apply to all broad-acre pesticides, but more localized uses such as rodenticides may not require these additional data. In practice, however, if the compound is rapidly lost from the aquatic environment such that the dissipation half-life (DT_{50}) is less than four days, then it as assumed that chronic exposure will not occur and chronic studies may not be required.

When repeated or continued exposure of the aquatic environment is predicted to occur, chronic studies on an aquatic insect species and an aquatic gastropod mollusc are also triggered. At present, no standardized guidelines for these studies are available. In practice, these studies are only likely to be required for products which are applied directly to water, *e.g.* aquatic herbicides. Sediment toxicity tests with *Chironomus* (a midge larva) may fulfil the insect testing requirement. Sediment toxicity testing is also required for compounds which adsorb to and persist in sediment and are potentially toxic to aquatic organisms (see Maund *et al.*[13] for more discussion).

A chronic fish toxicity study is required unless it can be justified that repeated or continued exposure of the compound to aquatic organisms will not occur, or unless suitable microcosm or mesocosm data on fish are available. There are two standard chronic fish toxicity tests in the EU: the extended mortality and early-life stage (ELS) tests. As their names imply, the former evaluates toxicity over 28 days, and the latter examines effects on the early-life stages of fish. The early-life stage test is generally considered to be a higher tier study than the extended mortality test. Bioconcentration studies with fish are required for compounds with a log P_{ow} (octanol: water partition coefficient) > 3, unless it can be justified that exposure leading to bioconcentration will not occur.

For registration of pesticides in the USA, further data requirements are described under The Federal Insecticide, Fungicide, Rodenticide Act (FIFRA) 40CFR Part 158. Additional species of algal and *Lemna* studies are required in practice for herbicides, or for compounds which show significant toxicity to algae (*Selenastrum*) at expected environmental concentrations. Chronic studies on

---

[13] S. J. Maund, I. Barber, J. Dulka, J. Gonzalez-Valero, M. Hamer, F. Heimbackh, M. Marshall, P. McCahon, H. Staudenmaier and D. Wustner, *Environ. Toxicol. Chem.*, 1997, **16**, 2590.

freshwater fish and *Daphnia* are essentially a basic requirement for US registration and are only not required if it can be justified that: continuous or recurrent exposure will not occur; acute toxicity of the compound is $> 1\,\mathrm{mg}\,l^{-1}$; expected environmental concentrations are more than 100 times below the toxicity to any aquatic organism tested; other data do not suggest that fish may be affected; the compound has no potential for cumulative effects; and the compound half-life in water is less than 4 days. Fish bioconcentration studies are also required for all compounds. However, a waiver may be granted if certain conditions apply, *e.g.* if the log P_{ow} is less than 3 and there is no evidence of accumulation in mammalian or avian species.

In addition to studies on freshwater organisms, for US registration, acute (96 h mortality) estuarine/marine fish and shrimp and 48 h oyster embryo larval studies are required for compounds intended for direct use in the estuarine or marine environment, or for compounds whose use pattern or mobility might lead to significant exposure in these environments. For example, a non-mobile compound whose use was limited to mid-western corn would not require these studies.

For Japanese registration, no testing of aquatic plants is currently required. If the 48 h LC_{50} for carp is less than $0.5\,\mathrm{mg}\,l^{-1}$, then tests on additional fish (including trout, loach and killifish) and invertebrate (shrimp and shellfish) species are often required. In addition, if the compound fulfils these criteria and the intended use includes rice, or the acute toxicity to *Daphnia* is less than $0.5\,\mathrm{mg}\,l^{-1}$, then a *Daphnia* chronic study should also be performed.

Additional Study Requirements for Pesticide Products

As well as studies of the active ingredient, certain studies are also required of pesticide commercial end-use products (EUP), which may include a range of co-formulants. Requirements for studies of the end-use product differ between the regulatory authorities, and require a certain amount of interpretation. In many cases, data generated on similar formulation types (*e.g.* two comparable emulsion concentrate formulations) can fulfil the requirement for additional formulation data because of the similarity of the product. Mostly, EUPs are not significantly different from the active ingredient in their toxicity.[14] If this is the case, data on the active ingredient are used for the risk assessment. Where the formulation is significantly more toxic, assessments are then based on formulation toxicity and further studies to evaluate the differences in toxicity may be required.

In principle, for EU registration the studies described in the basic data requirements above should be performed for each EUP. However, where data on the active ingredient indicates that one group (*i.e.* fish, *Daphnia* or algae) are particularly sensitive, tests can be limited to the most sensitive group. Studies of the EUP are also specifically required if it contains adjuvants or other active substances which could potentially increase the toxicity in comparison to the active ingredient, or its intended use includes direct application to water (*e.g.* aquatic herbicides). Further testing using higher tier studies may be required if

[14] R. Schmuck, W. Pflueger, R. Grau, U. Hollihn and R. Fischer, *Arch. Environ. Contam. Toxicol.*, 1994, **26**, 240.

there are significant differences between the toxicity of the EUP and that of the active ingredient. In principle, this could include any of the higher tier studies, depending on the area of uncertainty of the toxicity of the EUP. Chronic testing of formulations is not required on a routine basis.

In the USA, acute testing of EUPs is only required on a non-routine basis under the following circumstances: when the EUP is intended for direct application into the aquatic environment, when an ingredient in the EUP other than the active ingredient is expected to enhance the toxicity of the active ingredient, or when the expected residue level in the aquatic environment approximates any of the effect concentrations on the active ingredient. Where studies are triggered which require the use of an EUP, it is acceptable to perform the study of a representative formulation. Chronic testing is not required for EUPs, although any field studies are usually carried out using representative formulations.

In Japan, toxicity testing on carp and *Daphnia* is required for all EUPs. Toxicity of the EUP is then compared to a predicted concentration in the paddy, calculated from maximum use rates and a 5 cm depth of water. The ratio between this value and the formulation toxicity to fish and *Daphnia* is then derived. If there is less than a factor of ten between exposure and effect concentration for rice, or 20 for upland crops (using the same exposure calculation), tests on the additional Japanese species in Table 2 are required for the EUP.

Higher Tier and Field Studies

If preliminary risk characterization (see below) based on the laboratory toxicity data indicates that there may be concerns for a particular compound, further effects data may be required to demonstrate that there will not be adverse effects under field conditions (*i.e.* during normal agricultural use). Because standard laboratory studies are worst-case in nature (constant exposure in water alone), higher tier and field studies are usually conducted to incorporate mitigating factors which are not included in the preliminary assessment. These can include:

- Differences in toxicological responses—test species and life-stages for laboratory studies are selected for their sensitivity. In natural ecosystems, a wider range of potentially less sensitive organisms/life-stages may be present. In the field, behavioural responses may also allow the organism to escape chemical exposure. Laboratory studies are performed to give constant exposure during the test period, whereas in reality test concentrations are likely to fluctuate. For example, this may be important for hydrolytically unstable compounds where significant differences in observed impact can occur between dynamic laboratory studies (in which exposure is maintained at a constant level) and field assessments or laboratory tests conducted under static conditions (in which exposure concentrations are not maintained).
- Differences in the fate of the chemical—although the fate of the chemical is often well understood from laboratory studies, the combination of factors present in the field may lead to differences in exposure from that predicted in the laboratory. For example, if a chemical partitions rapidly into sediment,

laboratory tests in clean water may overestimate the potential impacts under field conditions.

- The ecology of the organisms of concern—although effects may be suggested by laboratory data, ecological processes of recovery may mitigate impacts under field conditions. For example, organisms with high intrinsic rates of increase may recover very rapidly under natural conditions, such that short-duration effects may be acceptable.

In order to identify a clear objective for this further effects evaluation, it is important to understand which of these factors are likely to be of importance for the compound under investigation. This then allows a targeted programme of experimental work to be defined to address the concerns raised by the preliminary evaluation. It should be noted that this may not necessarily involve outdoor or field studies. Customized, focused, laboratory studies may be more appropriate to evaluate the potential risks from a pesticide. For example, if it can be determined that an organism or group of organisms is particularly at risk from the compound, then laboratory studies on the impacts on populations of these organisms might be appropriate. Also, if it is known that a certain fate process (*e.g.* adsorption) may significantly reduce exposure, then laboratory studies in the presence of sediment or plant material may be able to demonstrate adequately that risks will be reduced under field conditions. However, if concerns are focused at the community level (*i.e.* a variety of taxa may be affected), or if concerns relate to potential impacts on function (perhaps a key component of the food-web is susceptible and secondary effects may be expected), then field studies may be the appropriate tool for evaluating potential impacts.

History of Aquatic Field Studies. Aquatic field studies with pesticides have been evolving over the last two decades. Initially, studies were performed in natural or semi-natural ecosystems, such as farm ponds or ditches. Applications of the chemical were made in crops neighbouring the water body, and monitoring of the ecosystem was carried out for a period after the chemical had been used. Whilst these studies benefited from a good deal of environmental realism, it was sometimes difficult to interpret the results of the studies because seasonal or other non-pesticide related changes could not be compared to a suitable control. Even though surrogate controls (*e.g.* another close water body in an untreated location) were used, there were still often substantial differences between the two systems. Without properly replicated systems, it was not possible to develop inferential statistics and, consequently, there were difficulties in basing regulatory decisions on these data.

As an alternative, replicated experimental systems were the next logical step, and whilst these had been used for many years as limnological experimental tools to evaluate impacts such as the effects of eutrophication, it was not until the early 1980s that studies began with pesticides. In the US, the test system that was required for registration was the pond 'mesocosm', a large experimental pond of at least $400\,m^2$ surface area, which was required to contain uncaged bluegill sunfish (a small, centrarchid sport-fish which feeds principally on larger zooplankton and insect larvae). These studies attempted to evaluate both fate and effects in the whole pond ecosystem, rather than targeting specific components for detailed study. Extensive measurements of pesticide residues in all matrices (water, sediment and biota) and a vast list of ecological measurements (sometimes measuring the same endpoint in different ways) were required. As a result, the objectives of the study were generally vague and the data diluted in an attempt to cover all aspects of the pond ecosystem. Added to this, without the presence of a top fish predator, bluegill populations often grew rapidly, sometimes resulting in more than 20 000 juvenile fish at the end of a study from a stock of 20 adults. This number of fish were a substantial imbalance in the ecosystem, leading to overpredation of many arthropod taxa, which were often key areas of interest in the study. As a result, the data produced from these studies were often difficult to interpret.[15] At a similar time in Europe, experimental pond studies to support pesticide registrations were carried out in much smaller systems, typically of around 20 m^3. These did not usually contain free-living fish, although caged fish were sometimes used as a bioassay within the study.

During the mid-to-late 1980s, a number of ecotoxicology researchers began to use much smaller outdoor systems (so-called microcosms) and it became apparent that there were a number of practical advantages to using these systems over the larger ponds.[16] As a result, in the early 1990s a number of international workshops took place to reach a consensus on the types of experimental systems that would be appropriate for studying fate and effects in microcosms and mesocosms. The two terms were subsequently redefined so that size is currently used to differentiate the two nomenclatures (microcosms are 1–15 m^3, mesocosms are > 15 m^3). Good technical guidance and methodological advice is now available for these studies.[17–19] In practice, most ecological endpoints can be readily addressed in microcosms, whereas larger ponds may be more appropriate for studies of fate.

Interpretation of Field Data. The data produced from mesocosms and microcosms are typical of ecological field data, with large data matrices containing counts of species, from a variety of treatments, on a number of sampling occasions. Analysis of the data should be determined by the objectives of the study.

One common objective in these studies is to determine whether there are significant effects on communities of organisms. In the past, this question was addressed by performing multiple comparison tests (such as Dunnett's test), comparing each treatment to the control on each sampling occasion. This generated a very large number of comparisons from which it was difficult to

[15] *Symposium on Aquatic Mesocosms in Ecotoxicology*, ed. N. O. Crossland and T. W. La Point, in *Environ. Toxicol. Chem.*, 1992, **11**, vol. 1.

[16] Society of Environmental Toxicology and Chemistry/Resolve, *Workshop on Aquatic Microcosms for Ecological Assessment of Pesticides*, SETAC, Pensacola, FL, 1992.

[17] Society of Environmental Toxicology and Chemistry—Europe, *Guidance Document on Testing Procedures for Pesticides in Freshwater Static Ecosystems*, SETAC-Europe, Brussels, 1992.

[18] *Freshwater Field Tests for Hazard Assessment of Chemicals*, ed. I. R. Hill, F. Heimbach, P. Leeuwangh and P. Matthiessen, Lewis, London, 1994.

[19] *Aquatic Mesocosm Studies in Ecological Risk Assessment*, ed. R. L. Graney, J. H. Kennedy and J. H. Rodgers, Jr., Lewis, London, 1994.

ascertain the overall effect on the community.[20] More recently, a number of new multivariate techniques have become readily available, which allow the data to be presented and visualized in a readily interpretable manner. These tests are inherently powerful because they enable all of the data to be considered together. They also permit data to be presented in an ecologically meaningful way.

Whilst the range of techniques for analysis of data is ever-improving, this does not mean that the scientist can abdicate responsibility for interpreting the results of these studies. Aquatic field data should always be interpreted with due consideration of the ecological relevance of the results. Without an understanding of the ecological relevance of the data, regulatory decisions based on mesocosm and microcosm studies can only ever be arbitrary, and it is the responsibility of scientists to ensure that the appropriate interpretation accompanies the studies. These are expert-designed and expert-conducted studies, so they should also be expert-interpreted studies. A number of ecological considerations should be applied when interpreting aquatic field data:

- *Recovery*—although impacts may be observed on certain organisms during a study, they may subsequently recover during the course of the experiment. Some consideration should be given to the duration of a difference before it is considered ecologically significant. This will vary for different types of organisms, *e.g.* high mortalities may be important for an organism with low reproductive rates and a long life-cycle, but may be of little significance to a fecund species with a short generation time
- *Relevance to other aquatic ecosystems*—measuring effects in one type of experimental system may provide useful information for determining what happens in that specific system. However, it may be difficult to interpret those effects in relation to other types of ecosystem. Differences in climate, size, location, *etc.*, may mean, however, that results from one study may not always be readily extrapolated to all other ecosystems. In many cases, mesocosms and microcosms provide an ecological worst-case because of their relatively small size and isolation
- *Spatial and temporal considerations*—mesocosm and microcosm studies are conducted under worst-case conditions, where the entire experimental system is treated with the chemical, often on a number of occasions. Application rates are based on assumptions of worst-case exposure, and combinations of these worst-cases may provide extreme conditions which are unlikely to occur in reality. Non-uniformity of exposure creates the potential for both spatial and temporal refugia in the real world which may not be present in field experiments

3 Exposure Assessment

Exposure assessment aims to determine the concentration of the pesticide in the appropriate compartment of the aquatic ecosystem for comparison with the effects concentration. It relies on the use of *exposure models* which combine the

---

[20] R. L. Graney, J. P. Giesy and J. R. Clark, in *Fundamentals of Aquatic Toxicology*, ed. G. M. Rand, Taylor and Francis, London, 1995, p. 231.

variety of factors which can modify exposure. A number of these models are available now and are widely used in pesticide risk assessment; these are discussed in more detail below.

Aquatic exposure concentrations of pesticides are determined principally by the following factors:

- The *use pattern* of the compound (determined by the label recommendations), including location, crop types, use rate and application method
- The *fate properties* of the chemical (*i.e.* its environmental behaviour before, during and after entry to the water body) dictated by its physico-chemical properties and degradability together with the use pattern determine the *route of entry*, the quantity of chemical which may enter a water body and its subsequent fate in the water body
- The properties of the *water body of concern* that may be exposed (in risk assessment this is usually based on some idealized form of water body), *e.g.* pH, sediment characteristics, depth, flow rate, size of watershed

Pesticide Use Patterns

The *use pattern* of the pesticide (application rate, timing and frequency of application, application method) is a critical first consideration in determining the potential exposure of aquatic ecosystems. Use patterns that are consistent with the area of concern for which a risk assessment is conducted should be applied. This is important because there can be major differences in the extent of exposure from different application methods. There are a number of key questions that should be addressed:

- Where is it used (topography, hydrology, soil characteristics, underlying geology and climate can make a big difference, particularly to pesticide runoff)?
- When is it used (what time of year, how often are applications repeated, occurrence of applications relative to seasonal high rainfall or recharge events in autumn and winter)?
- How is it used (are liquid applications sprayed by air, mist blower, ground hydraulic equipment (tractor sprayers); are granular applications broadcast across the soil surface or banded directly into the soil, incorporated by tines)?
- What crops are treated (are they tall crops, leafy crops with lots of interception)?
- How much is applied and does this vary with time of year or stage of the crop?

Pesticide Fate

As soon as a pesticide has been applied to its target crops, its fate properties begin to play an important role in determining the likely exposure concentrations. Immediately after application, factors such as the photolytic rate on leaves and soil, lipophilicity (affinity for the leaf surface), soil degradation rates and soil adsorption coefficients will be important in determining how much chemical is available to leave the target area. Once the chemical has entered the aquatic environment, a variety of fate processes also affect the concentration of the chemical in the water body, including degradation, partitioning and physical transport.

Degradation in surface waters can occur either as a result of chemical degradation processes such as hydrolysis and photolysis, or as a result of biological degradation processes associated with sediment and suspended sediment. The relative importance of each process in defining the fate of chemicals in surface waters depends largely upon the environmental conditions in the surface water body being investigated. For example, surface water pH conditions and temperature affect hydrolysis rates; turbidity and depth affect overall photolysis rates owing to differences in light penetration. The extent of biodegradation associated with sediment and suspended sediments depends not only upon the active microbial biomass present capable of metabolizing the compound in this environment, but also (in common with chemical degradation processes active in sediment and suspended sediment) the extent of partitioning occurring, which may transfer the compound from the free form in solution to the sorbed form.

Partitioning processes include volatilization, sorption onto and desorption from macrophytes, suspended sediments and sediments, and uptake by biota. As with degradation, environmental characteristics play an important role in the extent of partitioning. The predictability of adsorption and desorption processes depends upon the nature of the sorption processes (reversibility of sorption, correlation with organic carbon contents of sediment and suspended sediments and organic films associated with the surfaces of macrophytes). Generally, for most pesticides which are non-polar organic molecules, sorption is assumed to be related to organic carbon content and is assumed to be fully reversible. In reality, sorption processes are extremely complex and, particularly in dynamic surface water bodies, these assumptions may in some cases be quite incorrect. These processes can be extremely important in defining not only the profile of aquatic exposure with time but also the parallel sediment exposure time profile. The extent of uptake by biota depends largely upon the availability of chemicals (determined by their partitioning behaviour with sediment, plants, *etc.*), the life cycle, feeding strategy and lipid content of the organism in question, and the octanol–water partition coefficient of the chemical.

Physical transport processes include advection, sediment burial and atmospheric deposition (see below). Advection is simply transport of chemicals (either in the dissolved form or associated with suspended materials) in flowing water, and this is clearly related to the hydrological properties of the water body. Sediment burial generally occurs over long timescales and is, therefore, not often considered a significant loss process. However, in field situations where significant erosion can occur (susceptible soils, poor agricultural management practices, high slope, high rainfall intensity), residues can become buried under layers of deposited eroded material. The influence of bioturbation (organisms such as worms and insect larvae moving through the sediment) also can promote the downward movement of residues in sediment.

Routes of Entry

By combining knowledge of the use pattern and fate properties of the compound, it is possible to arrive at an initial view of the likely *routes of entry* of the compound into the aquatic ecosystem. These are described below.

Direct applications or overspray of pesticides to water is not consistent with Good Agricultural Practice, except for compounds such as aquatic herbicides or specialist biocides used in aquaculture where intentional applications into surface water are made. Although the use of an 'overspray' scenario was formerly regarded as a worst-case for spray drift exposure by some regulatory authorities (*e.g.* the accidental passing of a spray boom over a ditch bordering a field), it is now recognized that this is a relatively rare occurrence and in many countries is an illegal usage. Even for aerial applications, where arguably the potential for overspray is greatest, spray applicators go to great lengths to ensure the application is targeted on the crop to ensure efficient and economical pesticide usage.

Under certain conditions, off-target movement of spray applications—*spray drift*—can occur. The significance of this route of entry is highly dependent on the manner of application. For example, ground hydraulic applications with low boom heights with nozzles directed towards the ground tend to have lower drift rates than air-blast applications to orchards where the intention of the application is to generate a fine mist of chemical which will shroud the trees. Also the distance from the water body itself and the degree of development of foliage on the target plant can heavily influence the degree of spray drift. Owing to practical difficulties in routinely assessing potential for spray drift, it has become common to generate either look-up tables of drift rates for various crops at various distances (*e.g.* see Table 3 which is used in spray drift assessment for the EU[21]) or to develop empirical models of relationships between spray drift deposition and use pattern (*e.g.* 'AgDrift' which is being developed by the Spray Drift Task Force in the USA, or 'MOPED' in Germany[22]).

It is important that realistic scenarios for determining the extent of spray drift exposure are developed. For example, typical distances of crops from a water course need to be established for different crops, since distance is one of the key determinants in reducing spray drift. Whilst it may be appropriate to assume that the worst case for arable crops is that applications by ground equipment may take place as close as 1 m to a water body (*e.g.* for drainage ditches in Holland), this is less sensible for a mist blower application in an orchard, where 3 m (or even 10 m) may be a more appropriate distance. Also, the correct application of empirical data requires careful thought. Whilst assuming that the 95th percentile drift rate may be a reasonable worst case for a single spray drift event (*i.e.* 95% of the applications are described by drift rates equal to or less than a specified value[21]), combining several 95th percentiles for a multiple application soon results in a probability that is vanishingly small. In these cases, using the mean for each occasion except the last, where the 95th percentile is applied, may be more appropriate. Further consideration should also be given to the application of

[21] M. Ganzelmeier, H. Kopp, R. Spangenberg and M. Streloke, *Pflanzenschutz-Praxis*, 1993, **3**, 14.

[22] FOCUS, *Leaching Models and EU Registration*, FOCUS Group, EC Document 4952/VI/95, Commission of the European Communities, Brussels, 1995.

Table 3 Drift rates (% of application rate) for various crops at different distances from the treated area

Distance/ m	Arable/field crops	Vineyards		Orchards		Hops
		with leaves	without leaves	with leaves	without leaves	
1	4.0	—	—	—	—	—
2	1.6	—	—	—	—	—
3	1.0	7.5	4.9	15.5	29.6	—
5	0.6	4.0	1.6	10	20	11
10	0.3	1.5	0.4	4	11	7.5
15	0.2	0.7	0.2	2.5	6	4.5
20	0.1	0.4	0.1	1.5	4	3.5
30	0.1	0.2	0.1	0.6	2	2
40	—	0.2	0.1	0.4	0.4	0.6
50	—	0.2	0.1	0.2	0.2	0.3

interception from riparian vegetation, which may result in reductions in spray drift.

Surface runoff or *overland flow* occurs intermittently and its occurrence or location in a field situation is often difficult to predict.[23] Chemicals can be found in two forms in runoff: in the soluble form (associated with runoff water) and in the eroded form (sorbed to suspended soil). The proportion and quantity of a chemical in each form at a given site depends upon the extent of sorption with the associated soil matrix, partitioning between runoff water and the suspended eroded material in transit, and the degradation rate in soil. These are all, at least in part, defined by environmental conditions present at the site (temperature, precipitation intensity, soil moisture conditions, soil texture, soil organic carbon content, topography and agricultural practices at the site[24–27]). Runoff losses (expressed as a percentage of applied chemical) are also highly dependent upon field size, with losses increasing with decreasing field size.[25] Nonetheless, pesticides can be crudely classified as to their potential for runoff and erosion.[28]

Because of the complexity of the interacting factors which affect runoff, specialist studies are sometimes performed for compounds whose runoff risk is determined to be significant. Field study design aspects incorporating these assessments are discussed by Carter and Fogg.[23] However, the results of these exercises can be difficult to interpret owing to the stochastic nature of these processes. In addition, the conduct of these studies may not always be possible or

[23] A. D. Carter and P. Fogg, in *Pesticide Movement to Water*, ed. A. Walker, R. Allen, S. W. Bailey, A. M. Blair, C. D. Brown, P. Gunther, C. R. Leake and P. H. Nicholls, BCPC Monograph No. 62, British Crop Protection Council, Farnham, 1995, p. 71.

[24] G. L. Harris, in *Pesticide Movement to Water*, ed. A. Walker, R. Allen, S. W. Bailey, A. M. Blair, C. D. Brown, P. Gunther, C. R. Leake and P. H. Nicholls, BCPC Monograph No. 62, British Crop Protection Council, Farnham, 1995, p. 371.

[25] R. J. Jones, in *The Environmental Fate of Xenobiotics*, ed. A. A. M. Del Re, E. Capri, S. P. Evans and M. Trevisan, *Proceedings of the 10th Symposium on Pesticide Chemistry*, Castelnuovo Fogliani, Piacenza, Italy, 1996.

[26] R. A. Leonard, in *Pesticides in the Soil Environment: Processes, Impacts and Modelling*, ed. H. H. Cheng, SSA. Book Series No. 2, Soil Science Society of America, Madison, WI, 1990, p. 303.

[27] R. D. Wauchope and D. G. Decoursey, in *Research Methods in Weed Science*, ed. N. D. Camper, Southern Weed Society, 3rd edn., 1986, p. 135.

[28] D. W. Goss, *Weed Technol.*, 1992, **6**, 701.

practical. Soil modelling techniques (such as PRZM[29] or GLEAMS[30]) offer an alternative to assess the extent to which these loadings may be of concern. However, the relatively low general predictive accuracy of these modelling approaches requires that the user handles these simulations with great care.

In the US there is a significant regulatory focus on these processes as they can be major routes of entry into surface waters in certain regions, particularly those with high rainfall intensity and a propensity for soil erosion (*e.g.* certain mid-west states—sometimes also referred to as the Corn Belt). However, in Europe it is commonly assumed that runoff and erosion are comparatively minor routes of entry into surface waters, and are generally a localized problem which is controlled by agricultural practice, *e.g.* bunding and terracing to prevent soil losses.

As with spray drift, due consideration should be given to appropriate scenarios for runoff assessments. For example, recent research has shown that buffer strips of vegetation can intercept considerable amounts of surface runoff.[31,32] The use of such vegetative buffer strips is becoming more common in Europe but is already widely used in the US as an effective risk mitigation measure.

As well as surface runoff or erosion, under certain circumstances in agriculture it is necessary to install drains in order to control or lower a temporarily raised or permanently perched water table where there is a less permeable layer in sub-soils. In the event that relatively mobile compounds (high water solubility, low adsorption and relatively persistent compounds) reach these drains through leaching, such compounds can be introduced to surface water through *drain discharge* into ditches or streams. Drainage may also be an important route of entry for relatively immobile compounds if macropore flow conditions arise (*i.e.* cracking of clay soils in summer time). A related process known as interflow can also occur where lateral transport occurs within the perched water table where drains are not present. Because of the relatively complex nature of these processes it is not possible to estimate with any accuracy the extent of drainage loadings without the use of relatively sophisticated modelling approaches. One important modelling development which incorporates assessments of drain discharge is the MACRO model.[33]

Unlike other routes of entry where risk mitigation measures can be used effectively to reduce exposure, the effectiveness and (most importantly) practicality of risk mitigation measures associated with drainage remain relatively untested. A review of possible management approaches to minimize pesticide loadings to surface water by Harris[24] included the possibility of reducing such loadings through restriction of sub-surface drainage to periods where the water table was present in the soil profile. Although no single measure is likely to eliminate the occurrence of pesticides in water bodies, a combination of approaches may offer the long-term potential to reduce the risk of losses to surface waters.

[29] R. F. Carsel, C. N. Smith, L. A. Mulkey, J. D. Dean and P. Jowsie, *User's Manual for the Pesticide Root Zone Model (PRZM)*, Release 1, EPA-600/3-84-109, Office of Research and Development, US Environmental Protection Agency, Athens, GA, 1984.
[30] R. A. Leonard, W. G. Knisel and A. D. Still, *Trans. Am. Soc. Agric. Eng.*, 1987, **30**, 1403.
[31] L. Patty, R. Belamie, C. Guyot and F. Barciet, *Phytoma—La Défense des Végétaux*, 1994, **462**, 9.
[32] R. L. Jones, in *Proceedings of the Brighton Crop Protection Conference—Weeds*, 1993, vol. 3, p. 1275.
[33] N. J. Jarvis, *Ecol. Model.*, 1995, **81**, 97.

Atmospheric deposition (wet deposition of chemicals associated with precipitation alone or precipitation and aerosols, or dry deposition of chemicals associated with aerosols alone) has long been considered a relatively minor route of entry for surface waters. This can certainly be the case near the site of application in comparison with other routes of entry such as spray drift. However, as the distance from the site of application increases, the relative importance of this process increases dramatically. Whereas overspray, runoff and erosion obviously occur locally to the site of application and spray drift can occur over short distances, partitioning into air and aerosols can result in long-range transport of chemicals far from the site of application. Subsequent deposition may result in contamination of water bodies normally unassociated with agrochemical contamination owing to their relative isolation from agriculture. Although this contamination is, nonetheless, relatively minor in comparison with the contamination of water bodies bordering application sites, the nature of the chemicals susceptible to long-range transport and deposition [persistence and moderate volatility (*i.e. ca.* 1×10^{-4} to $1\,Pa$)] can result in significant and unexpected bioaccumulation of residues in unexpected locations.[34] However, the relative importance of this route of entry from a regulatory risk assessment standpoint remains an issue for research.

Water Body of Concern

The next step in arriving at the potential exposure of aquatic ecosystems is to develop a scenario for exposure based on typical or worst-case water bodies. Clearly there is a vast range of water bodies which may be located near to agricultural land, including small ponds, lakes, drainage ditches, canals, streams and rivers. As an initial worst-case, a relatively small, static water body is usually selected, since in such water bodies dilution will be low or non-existent and therefore exposure will be maximized. In the EU, the water bodies that are selected are typically farm drainage ditches with assumed depths varying between 25 and 100 cm. In the USA, the worst-case water body is the farm pond, which is assumed to be a 1 ha pond of 2 m depth surrounded by a catchment of 10 hectares of agricultural land.

More sophisticated scenarios are not yet available, although at present a significant amount of effort is being applied to the development of appropriate higher tier surface water scenarios for the various regions of Europe, sponsored by the FOCUS group of the EU Directorate General VI (Agriculture) which is responsible for the implementation of Directive 91/414. In the USA, effort is also being applied to identification of higher tier scenarios, particularly the development of regionalized exposure scenarios ('Tier III') and the identification of water bodies of significant ecological resource. Within wider environmental regulatory areas in the USA, more focus is being applied to the modelling of catchments, rather than individual water bodies.

[34] F. Wania and D. Mackay, *Environ. Sci. Technol.*, 1996, **30**, 390A.

Mathematical Models for Generating Exposure Concentrations

The sections above demonstrate that predicting the exposure concentrations for aquatic environments will be complex. A variety of mathematical models are available to synthesize some or all of these factors and to assist the risk assessor with these estimations. Modelling approaches can generally be split into preliminary screening models and higher tier models. Screening level models typically only include relatively simple environmental processes, a limited amount of information on the compound and are limited to a single or small number of worst-case scenarios. Higher tier models are more complex, include more information and are able to simulate a wider range of environmental scenarios.

Screening Models. Although the EU does not yet have a specific screening model, a standard scenario is used to calculate exposure concentrations. It is assumed for worst-case purposes that a shallow ditch (typically 30 cm) with no dilution from flow (*i.e.* static) is contaminated by spray drift, with a minimal buffer distance of 1 m for arable crops and 3 m for all other crops. Spray drift values are abstracted from the estimates of Ganzelmeier *et al.* (see Table 3). By combining this fixed input (worst-case spray drift) with a water body of fixed dimensions, the initial concentration present in the water body is readily calculated as follows:

Use pattern:	50 g active ingredient (ai) ha^{-1} in flowering orchard (with leaves)
% spray drift at 3 m:	15.5%
Drift deposition at 3 m:	7.75 g ai ha^{-1}
Water body dimensions:	30 cm deep, 1 m wide ditch of 100 m in length
Surface area:	100 m^2 = 0.01 ha
Water volume:	30 m^3 = 30 000 L
Amount deposited:	7.75 g ai ha^{-1} on 0.01 ha = 77.5 mg
Water concentration:	77.5 mg per 30 000 L = 2.6 μg L^{-1}

Initial concentrations can provide useful information, particularly for acute risk assessment purposes. However, since the chemical's behaviour in the water bodies may result in dramatic changes in both the concentration and relative quantities residing in sediment and water, it may be necessary to conduct more sophisticated modelling for some compounds (see *Higher Tier Models* below).

In the USA, *GENEEC*[35] was designed as a simple screening model to estimate aquatic concentrations from a few basic chemical parameters (soil water partition coefficient, degradation half-life) and the pesticide label information. The model crudely estimates the amount of runoff which might occur from a 10 ha field into a 1 ha by 2 m deep pond. The dimensions of the aquatic system have been designed to be consistent with the standard worst-case water body used (*i.e.* a farm pond) in the US EPA risk assessment process. The model includes estimation of the impact

[35] R. D. Parker and D. D. Rieder, *The Generic Expected Environmental Concentration Program, GENEEC, Part B, Users' Manual*, United States Environmental Protection Agency Office of Pesticide Programs, Washington, DC, 1995.

of agronomic practice (manner of application) on runoff and the quantity of chemical deposited on the surface of the water body as a result of spray drift. The model also considers both potential acute and chronic concentrations in the water body through estimates of the reduction in dissolved concentrations by adsorption onto sediment and degradation of the chemical within the water body. The model is designed to mimic crudely the capabilities of the more complex PRZM/EXAMS suite of models also designed by the US EPA and discussed below. Initial GENEEC calculations determine which chemicals warrant further, higher tier modelling.

Fugacity models have already been demonstrated as an effective screening tool for industrial chemical assessment in a number of exercises,[36,37] although they are used less widely for assessments of agrochemical fate. Fugacity models can be flexibly designed to allow simulations of the fate of chemicals following unsteady-state loadings to the environment, including processes such as degradation, volatilization, runoff, erosion, leaching, partitioning between surface water and sediment and suspended sediment, *etc*. The primary advantage of fugacity calculations for aquatic systems is that a considerable quantity of thermodynamic information is condensed in compact, elegant calculations which are easily interpreted and solved. Fugacity calculations and interpretations can provide new insights into contaminant behaviour, particularly in aquatic systems, as they can facilitate the quantitative expression of the diversity of processes that occurs in the environment.

SLOOT.BOX is a relatively simple model based on a single aquatic compartment and based on a Dutch ditch ('sloot'). Pulse loadings are simulated. Although sedimentation and resuspension are simulated, simulation of sediment concentrations is not carried out. Simulations according to SLOOT.BOX are a regulatory requirement in the Netherlands.

Higher Tier Models. *EXAMS* is possibly the best-known of the surface water models. The model is capable of simulating a number of aquatic compartments and water body segments, allowing investigation of impact downstream. Pulsed and continuous loadings can be simulated. EXAMS includes the capability of input compatibility with the PRZM model runoff and erosion output reports. Partitioning into sediment, suspended sediment and biota are simulated. PRZM/EXAMS includes the capability of simulating a wide range of degradation processes (hydrolysis, photolysis, oxidation, reduction and biodegradation). Volatilization losses are also assessed. Steady-state hydrology is simulated.[38]

WASP5 is another similar model developed by the US EPA with many of the same capabilities as EXAMS, including input compatibility with PRZM runoff and erosion output reports. Importantly, however, WASP allows the simulation

[36] D. Mackay, *Multimedia Environmental Models—The Fugacity Approach*, Lewis, Chelsea, MI, 1991.

[37] D. T. Jager, *Feasibility of Validating the Uniform System for the Evaluation of Substances (USES)*, Report No. 679102026, National Institute of Public Health and Environmental Protection (RIVM), Bilthoven, NL, 1995.

[38] L. A. Burns, D. Cline and R. R. Lassiter, *Exposure Analysis Modeling System (EXAMS): User Manual and System Documentation*, Report No. EPA-600/3-82-023, United States Environmental Protection Agency, Washington, DC, 1982.

of dynamic hydrology and, unlike most other models, allows the simulation of parent and metabolites.

TOXSWA is a recent model developed to supersede the SLOOT.BOX model. The model is capable of simulating a number of aquatic compartments and water body segments, allowing investigation of impact downstream. Pulse and continuous loadings can be simulated. Partitioning into sediment, suspended sediment and macrophytes are also simulated, as is steady-state hydrology.[39]

4   Risk Characterization

Preliminary Risk Characterization

Having developed assessments of exposure and effects, the next step is to perform preliminary characterizations of the potential risks of pesticides. In reality, tiered assessments of effects, exposure and hence risk are normally carried out contemporaneously, and have only been discussed here individually for the purposes of clarity.

As with exposure and effects estimates, risk assessment is also a tiered process. The lowest tier typically uses data from the standard toxicity tests and a screening level exposure model, which estimates worst-case exposures in a standard water body. Exposure and effect concentrations are then expressed as a ratio or quotient, and predetermined values ('triggers') of these expressions are used to determine whether the compound generates a level of concern and warrants further evaluation.

Under the EU Pesticide Directive 91/414/EEC, at present there is no standard approach to generating predicted environmental concentrations (PECs), beyond that which determines the initial loading of the chemical into the test system for a single application (described above). In this basic screening approach, there is no provision for the inclusion of dissipation or degradation through time, so assessment of chronic risks can be extremely conservative if the chemical disappears rapidly. Potential for effects is expressed by dividing the effect concentration by the exposure concentration to give a 'toxicity: exposure ratio' (TER). For fish and *Daphnia*, if the PEC > 1/100 of the acute EC/LC_{50} (acute TER < 100) or if PEC > 1/10 of the chronic NOEC (chronic TER < 1/10), then the chemical is considered to require further evaluation. For algae, if the PEC > 1/10 of the lowest EC_{50} (TER < 10), further evaluation is required.

Since this simplistic approach is of limited applicability to many pesticides (*i.e.* many are used more than once, have exposures which are significantly affected by their environmental fate) and many fail this screen, in practice a variety of different modelling approaches are used to estimate exposure more realistically. Agreement of an appropriate modelling approach is then normally the subject of discussion between the regulator and pesticide registrant. At present, discussions are underway to develop recommendations regarding appropriate aquatic exposure models and scenarios for the EU by the FOCUS group of the Agriculture Directorate (DGVI). If these further modelling efforts still indicate

---

[39] P. I. Adriaanse, *Pestic. Sci.*, 1997, **49**, 210.

that there is a level of concern for any of the taxa, then mitigation and/or higher tier assessments are required.

For preliminary risk characterization under US registration, GENEEC is used to determine exposure concentrations through time and these values are compared with data from the standard aquatic toxicity tests. Typically, averaged concentrations over similar exposure periods as the toxicity studies (*i.e.* 96 hours for fish acute, 21 days for invertebrate chronic, *etc.*) are used to derive 'risk quotients' (RQs). These RQs are then compared with standard trigger values to determine whether a level of concern (LOC) exists for that compound for a range of different endpoints. In simple terms, if the ratio of the exposure and effect concentration is less than 10 for acute effects (RQ < 0.1), or less than 1 (RQ < 1) for chronic effects, then an LOC is presumed. Different trigger values are also used for restricted use compounds and endangered species. Typically, if the compound fails these initial triggers, exposure estimates are refined using PRZM/EXAMS (see above). Failure of the risk assessment criteria after PRZM/EXAMS modelling then requires either label mitigation and/or further evaluation of the potential risks identified (see below).

Further Evaluation and Risk Management

If potential concerns are identified by the preliminary risk characterization, two courses of action can be taken:

- Either risk management measures can be used to reduce exposure by limiting use patterns (*e.g.* by lowering label use rates and numbers of applications, or by imposing no-spray zones between the use area and water courses) so that the potential for the chemical entering the environment is reduced. Exposure concentrations can then be recalculated with the lower inputs to re-evaluate the risks
- Or further evaluation or refinement of the risks is performed by using higher tier studies or modelling (see above). These studies tend to be developed on a case-by-case basis, depending on the particular issues of the product. This kind of work tends to have more in common with ecotoxicological research than standard regulatory studies

Risk management by exposure mitigation can be achieved in a number of ways. Label restrictions which limit uses of a compound (*i.e.* a compulsory buffer zone, restrictions on vulnerable soil types) can help to reduce surface water contamination *via* spray drift, drain discharges, runoff and erosion. A number of other labelling options may be applicable, such as reduction in application rate and number of applications, an increased application interval (so that there is less opportunity for build-up of residues, leading to lower runoff and drainage contributions), or restricted application timing (*e.g.* removal of autumn/winter applications that may lead to extensive leaching into drains) which can significantly reduce potential exposures. All of these must, of course, be balanced against the desired efficacy of the compound—reductions in application rate are of little help if the compound does not then work effectively.

S. Maund and N. Mackay

The inclusion of specific agronomic practices in the use of the compound can also mitigate exposure concentrations. For example, for granular applications, a label requirement for incorporation (*i.e.* covering with soil) after application can greatly reduce the amount of chemical available for surface runoff. Also, practices such as conservation tillage (to reduce soil erosion) and drainage on poorly drained soils can reduce runoff exposure. Planting of trees or hedgerows can also help to reduce off-target drift.

More recently, interest has focused on precision farming techniques, which allow for much more precise application of the pesticide to the target, with consequent decreases in off-target losses. Examples of this include laser operated mist blowers for citrus orchards, banders for soil pest control which turn off at the end of rows, and remote sensing of weather conditions that lead to diseases such as potato blight, enabling very localized applications. In the future, such high-technology solutions may offer a number of options for reducing pesticide exposures for compounds which may be of environmental concern. An international review of current procedures for reducing exposure of water bodies from pesticides has been conducted by the OECD.[40]

---

[40] Organization for Economic Cooperation and Development, *Activities to Reduce Pesticide Risks in OECD and Selected FAO Countries. Part I: Summary Report*, Environmental Health and Safety Publications Series on Pesticides No. 4, OECD, Paris, 1996.

Subject Index